A BOOK

OF

SIMPLE GARDENING

A small garden in Somerset

(*Five months before this garden was a carpenter's wood-yard*)

A BOOK

OF

SIMPLE GARDENING

ESPECIALLY ADAPTED FOR SCHOOLS

BY

DOROTHY LOWE

Cambridge :

at the University Press

1914

CAMBRIDGE
UNIVERSITY PRESS

University Printing House, Cambridge CB2 8BS, United Kingdom

Cambridge University Press is part of the University of Cambridge.

It furthers the University's mission by disseminating knowledge in the pursuit of education, learning and research at the highest international levels of excellence.

www.cambridge.org
Information on this title: www.cambridge.org/9781107544536

© Cambridge University Press 1914

First published 1914
First paperback edition 2015

A catalogue record for this publication is available from the British Library

ISBN 978-1-107-54453-6 Paperback

INTRODUCTION

The aim of a school garden is three-fold :

(1) To teach each child to garden.

(2) To produce an attractive garden for which the children are collectively responsible.

(3) To give a sound foundation of practical work and artistic effects, so that at the end of their school career a boy or girl should be capable of managing any garden for which they become responsible, either by their own work, or by gardeners under their orders.

This book has been written with these three objects in view, of which the last is the most important.

I also hope the book will be of use to anyone who wishes to take up gardening in later life, and who needs a simple book which treats the subject from the very beginning.

The most ordinary tools have been described, and the uses to which they are put explained. Blank pages have been inserted at the end of Chapters I, III, IV, VI, VII, VIII, IX, so that readers may add their own notes when desirable.

There is a simple descriptive chapter on soils and the laying out of gardens, which, although it stands at the beginning of the book, is perhaps chiefly of use to the gardening master or mistress and to elder pupils but its principles may be told even to children to explain why the garden is laid out in a certain style. Besides the children's gardens, the school garden should have a rock garden, rosery, shrubbery and wild garden.

By a shrubbery and wild garden, I mean not only a collection of shrubs in a shady part of the garden, which must be brightened as far as possible by a variety of plants that do well in the shade, but also a collection of some of the many beautiful flowering shrubs growing out of grass in which a succession of daffodils, bluebells, wood-anemones, etc. have been planted.

The older classes might with advantage have a certain section of the real garden to look after : each section should include a piece of mixed border, rose garden or wall roses, shrubbery and grass. I would not advise that girls should do the actual mowing, but rolling, edging and re-turfing both by sowing and sodding might well be under their care. The paths can be utilised to teach them the application of weed-killers.

I think if the older classes had a real responsibility of this kind, they would feel a keen interest in the work and there would not be a tendency to look upon it as an occupation for the little ones.

I shall be pleased to answer any questions with regard to insecticides, etc., and various methods and suggestions recommended in my book. Finally let me recommend every gardener to read Kipling's poem " The Glory of the Garden " (*A History of England*, by Kipling and Fletcher) which, owing to the difficulties of copyright, cannot be reprinted here, but which is, I consider, the best gardening poem ever written.

DOROTHY LOWE.

HINTON S. GEORGE,
CREWKERNE.
September, 1914.

CONTENTS

ILLUSTRATIONS

PLATES

TEXT-FIGURES

CHAPTER I

SOIL. MANURE. THE SITUATION OF THE GARDEN. LAYING IT OUT.
THE PATHS. LAYING OUT A ROSE-GARDEN.

Soil. The soil varies in different neighbourhoods ; some varieties
of soil most usually met with are sand, sandy loam, loam and clay.
Sand is a light soil but warm and poor, and needs much enriching,
after which many things do excellently in it.

Loam, which contains sand and clay, is the most satisfactory on
the whole, and grows things well.

Clay needs lightening but suits some things better than anything
else.

Natural farmyard and stable manure make a sandy soil closer and
more binding, while they lighten a clay one, so they may be added
to either with advantage.

There are also marl, a mixture of lime and clay, chalky soils and
peat. Peat if well drained is light in texture and most useful to mix
in a potting soil, and heaths, rhododendrons and azaleas thrive in
it—they loathe lime and chalk.

Manure. Soil is improved by all kinds of natural manures, by
decayed rubbish of weeds and garden refuse ; but if this latter is
going to be used in the garden, it is important that no roots of weeds
such as bindweed and celandine be thrown on the heap ; these and
seed-pods must be put in a separate heap and burnt.

Leaf mould is very valuable : the leaves should be swept up in
the autumn and put in a heap in the open air to decay ; evergreen
leaves should be thrown on the rubbish heap to be burnt as they are
useless for leaf mould.

Put any turf or grass edgings that are taken up, in a heap to
decay, as when decayed and mixed with leaf mould it makes the
best potting soil.

L. S. G.   1

Artificial manures are of great value but are rather expensive. It is often wisest to consult a local gardener as to the most suitable kind for the soil of the neighbourhood.

Choice of aspect.   In deciding where to place the garden if there is a choice of positions, it goes almost without saying that a southern or south-western aspect is desirable ; that is, the ground should slope in that direction, and not be shaded from the sun ; if possible, the ground should be sheltered from north and north-east winds. If it is not sheltered naturally, trees and shrubs, or fences that will grow quickly and well should be planted to break the force of the wind.

Laying out the garden.   When you begin to lay out the garden, arrange as far as possible that rows should run north and south, so that the sun shines evenly along their whole length. Make a careful plan on paper before beginning to lay it out. I will give a few suggestions which may be introduced into the plan according to circumstances.

A few hedges increase the appearance of size by the opportunity they offer for fresh beds and effects as one turns a corner, on the other hand they are apt to impoverish the soil, and plants will not grow well close up to them ; if you decide to have hedges, they must be planted with a purpose and not dabbed aimlessly down where they will not be really wanted ; also let them be of beautiful, not of coarse and ugly shrubs.

If there is a long narrow strip of ground the most effective way of using it is to lay out a good broad mixed border, say 4 feet wide, on either side of a central path : behind the border vegetables and fruit may be grown. There is no need to be afraid of vegetables and fruit or to think that they must be hidden. A good row of raspberries, red with fruit, is most decorative and attractive : and many other fruits and vegetables look very nice if well cared for —much better than big ugly hedges that grow right up to the mixed border.

On the other hand a formal garden laid out in stiff beds may look well fenced off ; often rose gardens are effective laid out in this way. For the surrounding fence of a rose garden a high yew hedge is the best, as its dark and sombre foliage makes a powerful contrast to the blaze of colour in the roses as they burst on the view.

A rock garden is more suitably screened off by the natural formation of the ground, aided by a variety of shrubs. A series of terraces, if the aspect is good, can be made to look charming and are a regular sun-bath for things that revel in a hot sun; many delicate plants will grow there that would fail utterly in an open situation. The terrace steps can be made beautiful, overgrown with alpine daisies and rock plants. If forest trees are to be planted it is necessary to think years ahead and to put them where, in their full-grown size, they will not overshadow or spoil a good piece of garden. If there is room, their beauty shews best on large lawns or in a woodland, where beneath them the ground should be carpeted with crocus, wood-anemones, bluebells and similar semi-wild flowers. At intervals clumps of azaleas and many other flowering shrubs may be planted.

While you are about it, take pains to get a really good selection of trees, there are many beautiful varieties that are hardly ever seen.

School gardens are apt to give a patchwork effect as a result of the individual work of each child. Instead of this a general effect should be aimed at; that is, the garden as a whole, not merely a number of well-kept patches, should be made the object of work.

For instance, suppose that an oblong patch be available, about 15 yards by 25 yards, and twenty gardens are wanted; each garden should be 3 yards by 5 yards.

Arrange the gardens side by side, five in a row in four rows (see plan). Let a nice path of gravel or flags, $2\frac{1}{2}$ feet to 3 feet wide, run between rows A and B, and C and D; and a narrow path of trodden earth or cinder, 1 foot 6 inches to 2 feet wide, run along each end of the whole plot and between B and C; and a narrow path, 1 foot to $1\frac{1}{2}$ feet wide, between each garden. The broad paths should run north and south if possible; 3 yards by 3 yards of each garden along the broad path should be devoted to flowers and the remaining 2 yards by 3 yards, facing the cinder path, should be used for vegetables and for a nursery garden.

Each of these plots can be in the care of one or more children.

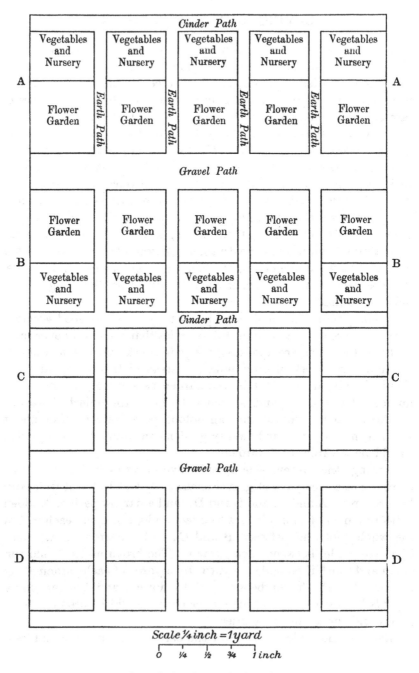

Ground plan for a school garden

Along one path the scheme may be that of a mixed flower border ; up the other path the effect of various masses of colour may be aimed at. The child's flower-garden should be made to look part of a long border, and yet at the same time it can be kept individually. The earth paths between the gardens should be as inconspicuous as possible.

SCHEME I. *Suggestions for a mixed border.*

At the back of the border a shrub or two (see Chap. VIII) and tall growing plants such as hollyhocks, helianthus, delphiniums, Michaelmas daisies, clumps of sweet peas can be planted ; in front of these come rosy larkspur, cornflower, love-in-a-mist, iris, poppies, etc., and quite at the front shorter plants, annual and perennial, such as dwarf marigold, phlox, godetia, violas, sweet Williams, etc. It is better to avoid a dead level of height, variety within moderate limits is more effective.

SCHEME II. *Massed colour effect.*

Let the children choose their favourite colours and then the gardens can be arranged either on a system of contrasts, or of a colour gradation. The two opposite gardens should be in each case of the same colour. Let us suppose yellow, red, blue; pink and mauve are chosen. The following are some of the plants that may be selected to carry out the scheme, in each case working from back to front.

Yellow. Helianthus, sunflower, dahlia, doronicum, marigolds (tall and dwarf), iris, eschscholtzia, daffodils, tulips, wallflowers, polyanthus, viola.

Red. Sweet pea, dahlia, rose, lychnis, poppies, phlox, lobelia cardinalis, red-hot poker, alonsoa, S. Brigid anemone (His Excellency), anemone fulgens, tulips.

Blue. Sweet pea, delphinium, borage (Dropmore), polemonium, cornflower, salvia patens, nemophila, lobelia, phacelia, hyacinths of various kinds, forget-me-not, commelyna, anagallis.

Pink. Sweet pea, dahlia, rose, rosy larkspur, phlox, clarkia, sweet William, dianthus (laciniatus), spirea, tulip, hyacinth, phlox Drummondi.

Mauve. Michaelmas daisy, sweet pea, phlox, scabious caucasica, veronicas (shrubs and plants), iris, ageratum, arctotis, viola, phlox divaricata, aubrietia, crocus (spring and autumn).

In the case of scheme I, I should suggest that the seeds should be raised by the class and divided later, each child taking a certain number of seed-pans under her special care.

In the case of Scheme II, the seeds of each colour should be raised by the owners of the two gardens of the same colour.

Several clumps of bulbs are advisable in the gardens : it is good experience to learn to hide the bare patch after they have died down.

Paths. If the broad paths can be laid with large flags rising their own thickness above the beds (or for this purpose large tiles can also be used) the effect is excellent, as low creeping things can be encouraged to grow out from the beds on to the paths, such as forget-me-not, sedum, nasturtium, saxifrage, viola, aubrietia and others, and, by breaking the line with splashes of colour greatly increase the effect from the end of the path. If the path is made of gravel or similar material a verge of simple tiles or low narrow lengths of wood is advisable ; the latter should be tarred or treated with creosote.

The rose garden. In front of the gardens might be laid out the class rose garden. The following design is suggested for either a square or oblong piece of ground.

Make a border all round having an entrance through the middle of each side, plant this with a single row of bush and standard or semi-standard roses : at intervals (10 feet to 12 feet) on the outside edge fix rustic stakes 8 feet high, joined by rustic poles and supported at the angles by diagonal pieces of wood. Up these stakes rambler and Wichuriana roses may be grown. Inside, 3 feet to 4 feet from the outer beds, opposite the entrances and at the corners, cut more rose beds (see plan), and plant them with bush roses, 4 feet apart. The paths and centre patch should be grass if possible ; brick tiling would also look well. Do not plant more than a single row of roses in each bed.

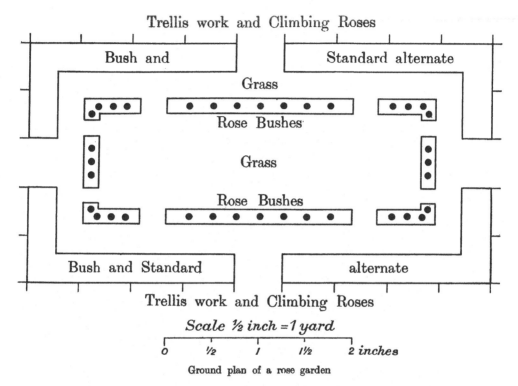

Ground plan of a rose garden

This shape of bed is really more effective than fancy beds, and easier to keep in order for mowing and edging. The roses being in a single row are easily reached and each bed can belong to two or more children.

NOTES

CHAPTER II

Tools. The following tools are necessary : a spade (medium size), four-pronged fork, Dutch hoe, rake, trowel, small hand or weeding fork, dibbler, pruning scissors.

Other useful tools are a second sized rake, draw hoe, small saw, and a mattock. In addition if there is turf, a pair of sheep-shears and a half-moon for cutting turves and edges.

Other desirable gardening implements are one or two sizes of sieves, a kneeling mat or board, garden-line, wheelbarrow and weeding baskets.

For a class of 12 girls I should recommend the following supply of tools :

2 spades	2 prs pruning scissors
2 forks	1 mattock
1 Dutch hoe	1 small saw
1 large and 1 small draw hoe	1 sieve
3 rakes	2 kneeling mats
4 trowels	1 wheel-barrow
4 hand-forks	3 weeding baskets
4 dibblers.	

Always wipe the soil off the tools before putting them away ; if they are dipped in a solution of soda and water you will find they will not rust for a long time, even if left out in the open.

The spade. To use the spade, hold the top of the handle with the right hand ; with the left grasp the shaft half-way down ; hold the spade upright and dig straight down into the soil by forcing the spade down with the arms and at the same time pressing with the left foot on to the left shoulder of the blade. Dig always a full spit, that is, as deep as the blade of the spade.

If you are digging a patch of ground, a trench should first be dug at one end and the soil wheeled to the other end ; then each spadeful of soil can be thrown into the trench in front of you, turning the under soil uppermost as you do it. The barrow load of soil from the first trench is used to fill up the last.

The four-pronged fork is used instead of the spade in sticky soil or in clay, or when the soil needs less deep stirring, or when there are many roots and plants in the bed.

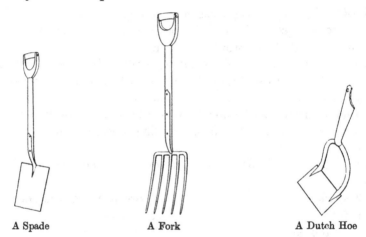

A Spade   A Fork   A Dutch Hoe

Push the fork in about two-thirds of the length of the prongs ; between plants it must be used so as not to disturb the roots unnecessarily ; it should also be used between roses and raspberries and any surface-rooting plants, or in ground full of weeds, of which the roots have to be picked out, as then it is easy to pull them out whole, the fork not cutting them like a spade.

If it is used for lifting potatoes the fork ought to have flat prongs, for other work, round prongs are better.

To dig potatoes insert the fork under the clumps of tubers, lift the clump and turn it over to the left ; the potatoes will then mostly lie on the top ; scatter the soil to let them dry and toss out with the prongs of the fork any that are partly buried, taking care not to pierce them.

Hoes. A Dutch hoe is a flat piece of steel about 6 inches by 2 inches, with a half-circle of metal which joins it to the handle.

Hoes and rakes are held in the same way, the right hand above the left about three-quarters way up the handle, where you feel you have full control over the tool.

The Dutch hoe is used for weeding and lightening the soil in dry weather but it is not suitable for heavy clay soil. Hold the hoe

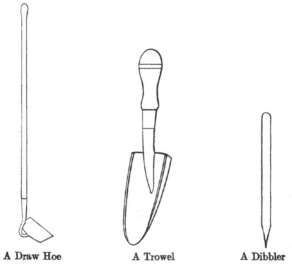

A Draw Hoe   A Trowel   A Dibbler

so that the blade lies almost flat on the ground and push it into the soil cutting about ½ inch below the surface for a space of 9 inches to 15 inches. You work walking backwards on to the unhoed ground, hoeing a fresh bit with each step back.

With a draw hoe you work walking forwards, treading on the hoed ground. Hold the hoe at such an angle that, when drawing it about 12 inches towards you, you will cut about an inch into the soil. It is a harder tool to use than the Dutch hoe, as if you hoe too deeply, it is very tiring and the hoe buries itself in the ground ; and if you hold it too lightly, it slides along the top and does no good and the stroke has to be repeated.

The way to handle a *trowel* needs no explanation. It is most useful for planting small things which need a larger hole than a dibbler makes and for filling pots and boxes, and uprooting small plants.

The dibbler, which is usually a piece of rake or broom handle 15 inches long and slightly pointed at one end, is used for pricking out seedlings ; and holes of various sizes can be made with it, according to the root of the plant. Make a hole, put in the seedling,

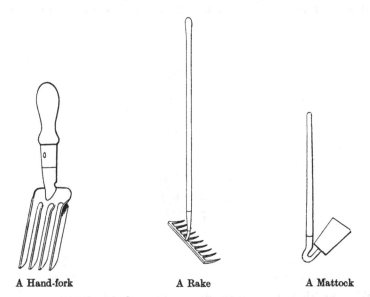

A Hand-fork   A Rake   A Mattock

holding it with the left hand, give a poke with the dibbler in a slanting direction beneath the seedling to push the soil round the ends of the roots. Then, with another poke or two from the top, push the soil into the hole and make the seedling firm, and the soil smooth and tidy round it. Be careful not to crush the rootlets nor to twist or curl them at the bottom of the hole. A dibbler should not be used when the ground is sticky.

The way to use a *hand-fork* again needs no description ; it is a small four-pronged tool the same size as a trowel, and is very useful for lifting weeds or for forking lightly in the flower-beds, when they are full of plants in the summer, and in the rock-garden where there is no room to use a large tool.

The rake is held like a hoe. Draw it towards you lightly and strongly. If it is not held firmly, it slides along the top or digs into the soil and makes a lumpy bed. It should be held at about the same angle as a draw hoe. Pull it towards you, then push back some of the soil with the rake and pull again and so on, working the soil till it is even and finely crumbled and the weeds are drawn to you in a heap as free from soil as possible ; often clods of earth must be broken by hitting them with the teeth or back of the rake. Remember to rake lightly and firmly or your bed will not be even.

A Forked Hoe or "Caxton"   Sheep-shears

The mattock is a narrow but very deep-bladed draw hoe ; the blade is about 8 inches long by 4 inches wide, and it is used to hack into hard soil or to make a trench by digging the blade into the soil and drawing it up with the soil which is then thrown to one side. It is used for making trenches in which to sow peas and beans, etc. and for earthing up potatoes and for the first earthing of celery.

The forked hoe or " Caxton " has the shape of a hoe, but has curved prongs of varying lengths instead of a blade : it is a very convenient tool in flower-beds or the rock-garden as it digs and rakes at the same time.

Sheep-shears are large spring-scissors used for clipping grass edges. Work moving forwards, and keep the back of the shears close to the turf edge while you clip, and cut well down to the path or bed so as to make sure that all the grass growing out of the side of the edge is cut.

A half-moon is a piece of steel of that shape with a handle and is used for cutting edges and turves. It cuts neat and clean and does not tear up the edges: be careful to cut down straight; the turf can then be lifted with a spade or proper turf-lifter, each turf being cut about 15 inches square and 2 inches deep, or rolled up as it is cut underneath with the spade in a roll 2 or 3 feet long and 12 inches wide.

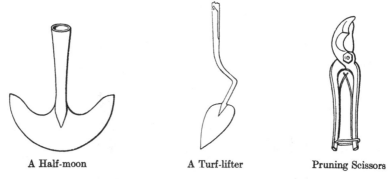

A Half-moon   A Turf-lifter   Pruning Scissors

Pruning scissors are strong scissors with a spring and are used for pruning, cutting off small boughs and cutting down dead plants. They are made in various patterns and a strong pair that opens wide should be chosen.

The sieves are used for sifting potting soil. Two sieves of different sizes are desirable; the larger one with a $\frac{5}{8}$ inch mesh for potting soil, and one with a $\frac{3}{8}$ inch mesh for preparing the soil in which to sow the finer seeds.

CHAPTER III

SOWING SEEDS. WEEDING. MULCHING. CUTTINGS. RUNNERS AND
LAYERS. POTTING. SYRINGING. PRUNING FRUIT TREES AND
SHRUBS.

Sowing in pans and boxes under glass. Crock the pan well by
putting bits of broken pot over the holes and bottom of the pan,
or fill the box a quarter full of rubble, *i.e.* the coarse remains that
will not go through the sieve ; then fill the box or pan to within
half an inch or an inch of the top with fine sieved soil, preferably
leaf mould, mixed with a handful of silver sand to a medium sized
box.

Press this down firmly and sow the seed thinly : if the seed is
very fine, mix it with a little sand and scatter it evenly over the
surface of the box ; cover it with a layer of soil as thick as the size
of the seed. The soil should be moist, not wet. Cover the box
with glass and do not put it in the blazing sun. Cover the glass,
or seed-box without glass, with brown paper or newspaper till the
seed has germinated. Keep the soil just damp ; seed-boxes must
be watered with a very fine rose.

If you have no heat at all, do not sow before the end of February
at the earliest, in the north still later. Cover the frame with mats
to keep the frost out. If you have a hot bed and frame, sowing can
be begun any time in February (see hot-bed, Chap. IV, April).
The temperature should be 75 degrees to 80 degrees and very little
air must be let in until the seeds are up, when gradually more should
be given. The frame should be closed again before night to keep
the warm air in.

Sowing out-of-doors. Fork up the patch where the seed is to be
sown, scatter over the place some finely sifted leaf mould, sow the
seed thinly and cover it with a thin layer of fine soil.

Some larger seeds as nasturtiums and peas should be sown deeper, the latter as much as two or three inches deep. Soak peas in paraffin for a minute and then dust them with red lead before sowing if there is any fear of field-mice. Certain slugicides are excellent preventives against slugs and do not hurt the seedlings; soot is apt to burn them and should only be put round the clumps, while these mixtures may be powdered over the seedlings as well, as they do not harm them.

If birds are troublesome, pea-guards may be used ; or black cotton tied to sticks and carried from one end of the row to the other, or strained across the clumps, keeps them and also rabbits away.

Pricking out. By this is meant planting out the seedlings a few inches apart either into a nursery bed, another box or into their flowering places. Naturally if they are going to be pricked out again they can be pricked out more closely than if they are only to be planted out once. Only experience can tell you the distance they should be apart, but rarely less than 4 inches to 6 inches in the open ground ; while in a box, for instance, lobelia may be planted 1½ inches apart.

Some things should not be pricked out, especially the plants with carrot-like roots such as eschscholtzia, poppy and cornflower. These must be sown in their permanent places and thinned out to a proper distance apart ; it is therefore a great waste of seed to sow them thickly. (Note. Autumn sown cornflower can be satisfactorily transplanted.)

Pricking out is usually done with a dibbler : see that the hole is large enough, and the seedlings planted firmly ; for larger seedlings use a trowel. They should be dug up carefully with a hand-fork out of the seed-pan.

Weeding. A weed is a plant growing in the wrong place. Seedlings especially must be kept free from weeds or they will be choked, as the weeds, being natural products of the soil, grow more strongly. Weeds that have runner-like roots such as bindweed and white-ash must be rooted out very carefully and every particle of the root removed, as any tiny bit will grow. Weeding should be done before the weeds come into flower, or seed. If the ground is dry

and the sun hot and the weeds small and not in flower, there is no
need to rake them up after hoeing, as they will die ; but if the
weather is uncertain it is better to take them up lest they root again.

Mulching is putting a layer of something round plants, for warmth
in winter, for coolness and moisture in summer, or to feed the plant.
It is usually strawy manure as that combines feeding and protection,
but if only warmth and moisture are required other things can be
used such as hay, straw, cocoanut fibre, grass cuttings, ashes and
leaves.

Cuttings. The three kinds are :

(*a*) A slip or sprout 3 inches to
5 inches long cut off close below a bud
either straight across or slantwise,
taking care not to cut into the bud.
If the cut is made slantwise, cut downwards and across, just below the bud.
Two or three pairs of leaves are then
cut off above the bud and it is ready
to strike. Very little more than the
top cluster of leaves in geranium or
chrysanthemum cuttings should be left.

(*b*) The same sort of piece, but
pulled off the old plant with what is
called a heel of the old wood ; that is,
it is pulled off at the bud where it
sprouts from last year's wood, and a

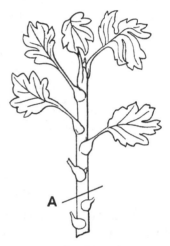

A cutting
A. Cut below a bud

little bit about ¼ inch or ½ inch long of last year's growth is torn off
with it.   Usually this method is used for roses and shrubs.

(*c*) A piece of the old plant with some rootlets attached to it :
these naturally root the quickest.

To strike cuttings fill a pot with soil, having crocked it well :
insert the cuttings either in the centre or pressed firmly against the
side of the pot : press the soil firmly down and keep it on the dry
side ; several cuttings may be put in one pot if it is sufficiently large.
Soft-wooded cuttings like chrysanthemums root quicker if kept from
the air under a bell-glass, or if the pots are put into a box deep enough

to hold pot and cutting and covered over with a sheet of glass. Most seedlings in boxes and cuttings should be put into a frame or cool house. If you have no heat at all, do not take chrysanthemum cuttings till February or March. Cuttings of fruit-trees and shrubs should be inserted 6 to 15 inches in the ground, and left for a year without moving.

Runners and layers. Runners are little plants forming at the end of a long stem, or rooting at intervals along a creeping stem—

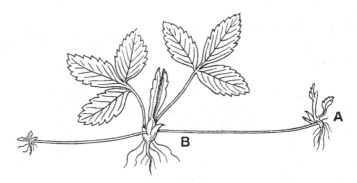

A Strawberry runner
A.   Parent plant   B.   Rooted runner

e.g. strawberries and violets. The best way to grow plants from runners is to peg the runner into the soil or into a pot sunk in the ground, covering the joint from which the leaves are growing until it is rooted ; then cut it off and leave it for a few days undisturbed, and after that pot it.

Layers are stems of plants pegged into the ground and encouraged to root. Some plants layer themselves and the layers have only to be taken up and planted ; others, for example some kinds of shrubs, will root if the bark is scraped with a knife and then pegged into the ground and weighted with a good sized stone to keep it in position.

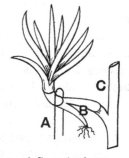

A Carnation layer
A. Pin to fasten the layer into the ground
B. Open cut in the stem of the layer
C. Stem of the parent plant

Others, like carnations, need the stem to be cut half through in a slanting upward direction immediately below a joint and to be pegged into the ground, keeping the cut open.

Carnations ought to be layered in July, before or as soon as they have finished flowering. As soon as they are well rooted they should be cut off the parent plant, and transplanted to a bed or cold frame at the end of September, or else left where they have been layered till March ; this latter course renders them less liable to damp off ; that is, rot away through excessive moisture.

Potting. Pots should always be clean—scrub them inside and out with water or soda and water ; new pots should be allowed to soak for an hour before using. Crock them well by putting bits of broken pot over the hole and round it, then, in medium sized and large pots, fill a quarter way up the pot with coarse rubbish that will not go through the sieve, or fibrous roots of decayed turf ; then fill up with sieved soil consisting of leaf mould, silver sand, decayed turf and sometimes wood ashes, till the pot will take the plant nicely ; place soil round the roots pressing it all firmly, but not hard, with the fingers, not with the thumbs.   Fill the pot nearly to the top, leaving sufficient space for watering ; tap the pot to make it all level and water with a rose for a day or two till the soil solidifies.   Bulbs need the soil beneath them to be light and hardly pressed down at all, but the soil above them should be firm.

Syringing should be done with a spray syringe if an insecticide is being used as it makes a mist-like spray and penetrates better into the plants, while it economises insecticide. Syringing with clear water should be done in early morning or evening in hot weather under glass. The following are some of the most common remedies applied by a syringe.

For green fly use any of the best insecticides on the market or one may be made of paraffin, soft soap and water (see Chap iv, July).

For mildew use a sulphur mixture.

For moss and green on trees use a lime mixture. Put a little quicklime into a bucket of water and use it when it is cold and has settled.

If chrysanthemum tips and other shoots are infected with the black or brown fly, use a similar mixture as for green fly. Put some of the mixture in a small basin and wash the shoot carefully with a small piece of sponge. After using any of these insecticides, the next day the plants should be syringed with clear water. Several applications of insecticides are often necessary, and are less harmful than one strong application.

Pruning. There are two different types of pruning : one causing the new, and the other the old wood to fruit.

(a) *To make the new wood fruit.* Cut out completely any weakly old wood which has no good young shoots growing from it, and occasionally even if it has, if there is too much new growth in the tree. Then cut a few inches off the tips of the young shoots to make them a convenient length. In bushes encourage as much as possible growth sprouting from near the ground, as this is more easily trained into a good shape. Cut back the old wood to just above a new strong shoot. When strong new growth is made from below the ground as in rambler roses, cut out the old wood completely. Most flowering shrubs, Wichuriana and rambler roses, black currants, gooseberries, and raspberries blossom on the new wood ; also peaches, but they require less hard pruning.

Many flowering shrubs require little more than cutting out old and weakly wood and cutting off the tip of the young wood if it is too long and poor ; standard trees are treated in the same manner.

Yellow jasmine has the flowering wood cut back to an eye or two from its base as soon as it has finished flowering and from there it will make its growth for next season. Most flowering shrubs should be pruned immediately after flowering.

(b) *To make trees fruit on the old wood.* Cut back this year's wood to three or four buds from its base ; this throws all the strength into the old wood below, and encourages it to make fruit buds. If the size of the tree is to be increased, leave a few leading shoots of this year's wood about 8 inches or 10 inches long. Nail or tie in the new wood on wall trees and thin out weak wood where necessary.

Summer pruning is cutting off 8 inches or 9 inches of the super-fluous growth of new wood in August or September. It is done to allow the sun and air to get freely into the tree to ripen the wood for next year's fruiting, and, by shortening them, to strengthen the shoots that have been cut back. Not less than 10 inches should be left of each shoot as they are liable to sprout again, and if they are cut hard back they sprout from the buds at the base of the shoot which should be preserved for next year's fruit blossom.

NOTES

CHAPTER IV

WORK FOR EACH MONTH.   SUGGESTIONS FOR WEEKLY CLASSES.

JANUARY.

Outdoors.   Any fruit pruning that was not finished in the autumn should be done as soon as possible except peaches, nectarines and raspberries.

Roll lawns once a week and mow if necessary : this last is not likely to be needed unless the winter has been very mild.

Finish digging up and manuring all the empty ground.

Any alterations of shrubberies may be made ; rose beds should be prepared for February planting (see Chap. VII). Trees, bushes and hardy creepers may be planted if the weather is mild and the ground not too wet.

Prune and train Virginia creeper, cutting back unnecessary and weak growth. Prune honeysuckle by shortening last year's shoots to 2 inches from their base.

A good supply of leaf mould and sand must be got ready for seed sowing.

Indoors.   Keep the last chrysanthemums free from green fly and black fly. Cut down those that have flowered ; this will encourage them to produce cuttings. Take bulbs that are well rooted and have about an inch of growth, from under leaves or out of the dark room in which they have been rooting and bring them gently into full light ; do not put them into strong sunlight for a day or two or the leaves may get burnt. If the sun is bright put them under the stage in a greenhouse or in a north room for a day or two, or even in a light outhouse if they will not get frosted.

A supply of clean and empty seed-pans and boxes must be got ready and sticks and labels prepared for the names of the seedlings. Rhubarb may be forced by lifting a large clump and putting it in the ground in the dark under a stage in the greenhouse or in a dark warm cellar or outhouse.

The seed list should be made out and the seeds bought as some will be wanted at the beginning of February if any heat is available.

Hints for weekly lessons.

First week. Prune fruit-trees. Dig. Roll. Prepare soil for seed-sowing. Attend to late chrysanthemums and bulbs.

Second week. Prune creepers. Dig. Roll. Force rhubarb. Attend to bulbs. Prepare seed list.

Third week. Prepare rose-beds for spring planting. Finish fruit-pruning. Study and prepare seed lists. Prepare seed-boxes.

Fourth week. Finish digging. Plant if the weather is suitable. Alter shrubberies. Roll. Prepare labels. Practise drawing up seed lists or lists for flower beds or shrubberies. Look at late pots of bulbs, and bring them to the light when ready.

FEBRUARY.

Outdoors. Lawns may be re-turfed and dressed. A good dressing for a tennis court is a cart load of loam and rotted manure and three bushels of soot and three of lime well mixed together and scattered over the surface. It should be raked off again after a month. If lawns are mossy, rake them well to scratch up the moss and dress them with wood ashes. Roll every week and mow if necessary but not when frost is on the ground.

Weedy and mossy loose gravel paths should be raked and weeded. Cut out the old wood from any rambler roses that were not pruned in the autumn and look over all trained trees and climbers in the fruit and flower garden ; nail or tie up where necessary.

Prune clematis Jackmanni by cutting back the shoots to a third of their length or by cutting the plant down to 18 inches from the ground. Some flowering shrubs may be pruned now. The purple buddleia may have its long stems cut back to 3 or 4 eyes ; if the tree is to grow larger leave about 2 feet of growth. Ceanothus (Gloire de Versailles) may be treated in the same way.

Cut back jasmine to three or four eyes as soon as it has finished flowering.

Plant roses, shrubs and hardy perennials if the weather is mild. In planting shrubs it is usually a safe rule to plant them as deep as the old mark on the stem which shews where they were planted in the nursery garden ; in any case see that the topmost roots are at least an inch below the ground. Roses and grafted shrubs should have the graft covered 2 or 3 inches.

Perennials that lose their leaves should have the crown just covered, those that have leaves near the base, as campanulas and primroses, should be planted as deep as the lowest leaves. German iris should be planted almost on the surface and have the tuber only half buried. All plants with a tap root should be planted in a deep hole and the soil rammed firmly round the root. Plant everything firmly and in a sufficiently large hole to take the roots without crushing them ; if anything comes up with a gentle pull it is too loosely planted. Re-plant montbretia.

Old growth from trees and shrubs should be cut away, and, if moss or lichen is growing on them, they should be syringed with lime water or some other suitable wash.

In a sheltered garden the first sowing of peas may be made. Dig a deep drill and work some manure in well and fill the trench to within 4 inches of the top. It is still better to have prepared this ground in November. Sow the peas thinly 4 inches deep with a dibbler ; if field-mice are troublesome soak the peas for a minute in paraffin and dust them over with red lead ; cover the drill with pea-guards or cotton strained from four or five sticks at either end.

Spinach may be sown between the ranks of peas if sufficient room has been left, i.e. not less than 6 feet between the rows.

Broad beans may be sown towards the end of the month. Sow in rows 3 feet apart, putting in the beans singly in a double row 4 inches deep and 6 inches apart each way.

The onion bed should be prepared ; the ground should be raked fine and then rolled (except on heavy soils) so that the seed may take a good grip. Sow a few early onions.

Plant Jerusalem artichokes using some of last year's stock or buying some from a greengrocer. Plant them singly about 18 inches apart each way, and 4 inches deep.

Force rhubarb and seakale out-of-doors by covering the clumps with a box or pot and heaping manure over it.

Put tow or twine black cotton over gooseberry bushes to prevent the birds taking the young buds. This may be necessary even in January. Towards the end of February prune peaches and nectarines by cutting out the old wood and cutting back a few inches of the new wood, always cutting *above* a leaf bud.

Raspberries should be cut back 4 to 8 inches, cutting away their weak or dead tops. If they are new canes cut them back to a foot from the ground.

Indoors. Make out the seed list at once if it is not already done. Bring out the remaining pots of bulbs into the light as soon as they are ready. Re-pot ferns. Sow lobelia and sweet peas in pots or boxes. In the third week sow lettuce, sprouts, early greens and tomatoes. Take chrysanthemum cuttings. If possible take a cutting with some root attached springing from the base of the old plant. If the cutting has no root, cut it square just below a joint, trim off two or three pairs of leaves at the bottom, put it in a pot of sand and leaf mould and press it firmly against the side of the pot and press the soil firm all round. Three to five cuttings may be put in a pot according to size—three in a 4 inch pot, four or five in a 5 inch pot. Label them carefully. Put the pots in a rather deep box and cover it over with a sheet of glass for two to four weeks as the exclusion of air prevents them flagging and helps them to root.

Hints for weekly lessons.

First week. Attend to bulbs and lawns. Prune shrubs. Prepare soil for peas. Put cotton on gooseberries. Make final seed list.

Second week. Sow lobelia indoors, peas outdoors. Plant roses and shrubs. Prune raspberries. Prepare potting soil. Pot ferns.

Third week. Sow sweet peas and vegetables indoors. Broad beans outdoors. Take chrysanthemum cuttings. Force rhubarb and seakale.

Fourth week. Sow seeds indoors and outdoors. Plant perennials. Take chrysanthemum cuttings. Attend to indoor bulbs.

Outdoors. Plant any perennials and small bits of plants that have been cut off. Plant out seedlings that have been in a nursery bed through the winter.

Replant outdoor chrysanthemums, either by dividing the old plants or by planting young rooted cuttings.

Fill up the carnation bed with seedlings or rooted layers ; they prefer a mixture of road grit, sand and leaf mould.

Look round ties and trained plants after storms ; tie up where necessary. Towards the end of the month prune sheltered rose-beds and climbing roses. The mixed borders should be weeded and raked ; tall nasturtiums and hardy annuals may be sown out-of-doors in mild climates.

Put in slips of lavender in a sunny position a foot apart each way ; the slips should be a piece pulled off the old plant with a heel of old wood attached. Some gardeners like to leave the slips lying on the ground under a hedge for 2 or 3 weeks before planting.

Sow sweet peas, soaking them in paraffin and dusting with red lead if there are field-mice.

Montbretia may still be re-planted.

Roses may be planted at the beginning of the month.

Dahlias may be started ; put them in a frame or in heat ; cover them with soil and water slightly.

Go through the rockery carefully, divide and re-plant where necessary ; fill up gaps with new plants or plants from the nursery bed which should have been begun last May. Rosettes may be pulled off saxifrages and planted singly. Plant gladiolus.

Roll and mow the grass regularly and trim the edges.

Peas and broad beans must be sown. Plant early potatoes 4 to 6 inches deep. Sow parsley in drills a foot apart and an inch deep—it should be sown very thinly.

Beetroot, early carrots, parsnip, lettuce and spinach may all be sown. In the third week sow the main crop of onions.

Globe artichokes may be increased by cutting off and planting, 2 feet apart, the young crowns growing round the old stem ; cut them off with as much root as possible.

At the end of the month plant the second early and main crop potatoes. Stick young peas as soon as they are ready. Flat hoe, that is, draw up the earth round the young potatoes as soon as they are 4 inches high. Clear the mulching from asparagus beds, and dress them with a good powdering of common salt.

Shade peach trees with curtains till the blossom is thoroughly set ; they should also have a glass shelter projecting about 18 inches from the wall, if fruit is to be relied on.

Indoors. Sow celery and schizanthus at the beginning of the month. Later sow most of the annuals and perennials.

Continue to take cuttings of indoor and outdoor chrysanthemums. Begin to dry off the bulbs in pots in readiness to plant out-of-doors in the early summer or autumn.

Nip down the sweet peas when they are 5 inches high if thin or drawn, and harden them off about the end of the month.

Prick out lobelia into thumb pots (the smallest size), and boxes as soon as it is ready to handle, putting three to six seedlings in each little clump. Other early seedlings may be pricked out into boxes. Re-pot geraniums and fuchsias. Plant lilies such as auratum and speciosum for late flowering.

Cuttings of geraniums, fuchsias and marguerites may be taken. Begonias should be put in boxes of cocoanut fibre and placed in gentle heat to start.

Hints for weekly lessons.

First week. Prick out lobelia. Dry off bulbs. Sow celery, etc. and vegetables indoors. Roll and trim edges. Tidy the beds. Plant montbretia. Take chrysanthemum cuttings.

Second week. Plant indoor lilies. Prick out seedlings. Sow annuals indoors, vegetables outdoors. Plant perennials. Start dahlias. Plant potatoes.

Third week. Pot geraniums, etc. and take cuttings. Sow annuals and perennials indoors, and sweet peas outside.   Attend to asparagus beds.   Divide globe artichokes.   Prune roses.   Plant carnations.

Fourth week.   Start begonias.   Pot geraniums.   Stick early peas.   Plant potatoes.   Weed and clear rockery.   Sow hardy annuals outside.   Prune roses.   Take outdoor chrysanthemum cuttings.   Prick out seedlings.

April.

Outdoors.   Carnations, sweet Williams, hollyhocks, Canterbury bells may still be planted.   Overgrown phloxes and Michaelmas daisies can be divided at the beginning of the month, also alpines may be planted out on to the rockery, and violas, pansies and hardy ferns be transplanted, the two last prefer shade.

Hardy biennial and perennial seedlings may be planted out if large enough.   Delphiniums may be increased by young shoots being cut off and potted up, pressing them firmly against the side of the pot ; they should be put in a frame or sheltered place.   Phloxes may be treated in the same way.   Polyanthus may be taken up when they have finished flowering, divided and planted in a nursery garden.

Plant out dahlias as soon as their shoots are 2 to 4 inches high, and towards the end of the month plant even those that have not started ; to increase them, cut them in two or three pieces, or cut off a young shoot with a bit of tuber attached and pot it up and grow it under glass for two or three weeks, and then plant out.

About the third week of April calceolarias may be taken out of the cold frame where they have been during the winter, and planted in a shady or semi-shady position after they have been hardened out-of-doors for a few days.

Harden off lobelia, geraniums, marguerites at the end of the month and other seedlings that are large enough to be pricked out, by putting them in a sheltered position or cold frame for a few days. Lobelia and geraniums require hardening for a week or two before planting them out at the beginning of May.

Sow seeds in open borders, forking and raking the ground and covering it with a layer of finely sifted leaf mould.   Scatter the seeds

thinly and cover them with a thin sprinkling of the prepared soil,
just covering the small seeds; larger seeds should be covered about
half an inch or an inch; press them down gently and if birds are
troublesome, put in a few twigs and twine black cotton among them
across the seed patch.   If there are many slugs dust the patch with
a slugicide or put a mixture of lime and soot *round* the patch as soon
as the seeds begin to germinate.   Begin to tie up oriental poppies
and paeonies if the weather is stormy; tie them loosely before it
is really required and tighten the string when necessary.

Pick dead flowers off the bulbs.

Laurestinus should have its dead flowers cut off, and Forsythia
and chimonanthes should have the branches which have flowered,
shortened.   Prune tea and noisette roses, cutting them back to
four or six eyes of last year's wood and cutting away all dead and
weakly wood.   Keep the beds hoed and weeded.   Plant out sweet
peas brought on under glass after hardening them for four or five
days by putting the pans or boxes out-of-doors or in a cold frame;
they should be sheltered for the first night or two if it is frosty or the
wind is cold.   Do not transplant them during a spell of cold windy
weather.   Dig the ground deeply and add manure if it was not
manured in the autumn.   Some gardeners dig as deep as three feet
but most are content with one foot; if the peas are to be in a row
make a trench 3 inches deep and prick the seedlings out 3 or 4 inches
apart in a single row.   Put small twiggy sticks for them to climb
up and stick them properly as soon as they begin to grow.   To
stick them get a bundle of pea-sticks 5 to 6 feet long and push them
into the ground on either side of the plants from 8 to 18 inches apart
according to their size.   Let all the sticks slope slightly in the same
direction and lean slightly inwards to interlace over the peas.   Sweet
peas may also be grown in clumps 3 feet across, the seedlings being
about 6 inches apart, when they can be grown up sticks, wire netting
supported on stakes, or twine netting circles.   Twine or wire
netting may also be used for rows of sweet peas, but only in places
sheltered from strong winds as it moves in the wind, which prevents
the seedlings getting a firm hold.

Take up old violet plants, divide them, pulling off young crowns with a little root attached if possible, or else runners with roots ; those that are to be put in frames in the autumn should be planted in semi-shade but the permanent bed must be where it will get full sun when the sun is low in winter time ; plant them in well dug and manured ground and never in the same place two years running. If necessary they must be watered till they are well established ; syringe them frequently in dry weather if there is any sign of red spider which shews itself by brown spots on the leaves. Keep all the runners cut off the young plants ; they should flower from the centre crowns and not from the runners.

Plant out bulbs that have flowered indoors or let them dry off in their pots. Continue to roll and mow lawns through the summer. Continue sowing peas at fortnightly intervals and stick them as soon as they require it.

The last sowing of broad beans should be made in the first or second week of the month. Scarlet runners should be sown about the second week, dwarf French beans in the third and fourth, and sowings continued at fortnightly intervals to ensure succession of crops.

Lettuces and early greens must be pricked out as soon as they are ready. Cabbages and other green crops should be sown, also annual herbs. In dry climates where the rain in early summer is slight, sow lettuce thinly broadcast so that it will not need pricking out.

Clear away the mulching from around the forced rhubarb towards the end of the month. Sprouting broccoli and other greens should be cleared away as soon as they are over and the ground dug and manured in readiness for peas or other late crops.

Salsafy, beetroot and carrots must be sown in drills if not already done—the last like a sandy soil well broken up and then made firm. Main crop potatoes must be planted at once. Cucumbers and marrows may be planted at the end of the month. Make the hot bed for cucumbers in a sunny position. Make a heap 4 or 5 feet high, or fill a pit and frame well above the level of the top of the frame with well-rotted manure that has been turned over two or three times ; it will soon settle down into a smaller heap and in

about a fortnight the temperature will have decreased sufficiently
for planting. The glass should be kept on the frame after the first
two days with an inch or two left open for the rank heat to escape.
After a fortnight, test the heat of the bed by pushing in a stick for
half an hour; when it can be held comfortably immediately on being
taken out the temperature is about right, *i.e.* 75 degrees. Make a
small heap of leaf mould in the centre of the frame in which to plant
the cucumber plant which has previously been raised in heat or
bought, and press the soil firmly round its roots.

The young plant should have four or six leaves besides its seed
leaves and the tip nipped off to encourage lateral growth.

To raise cucumbers from seed, put the seed singly in pots of
sand and leaf mould and place them in a temperature of 80 degrees.
In about a week after planting, water the plant with tepid water,
and give a slight syringing on bright days ; the frame should be
always left open an inch to allow the rank steam to escape. Cover
the lights with matting or straw if there are frosts at night.

As the roots push out through the heap of leaf mould cover them
up with soil 2 inches deep till the whole surface of the bed is covered.
The bed should then be thoroughly soaked once a week with water
of the same temperature as the soil ; each evening at 4 p.m. the
plant should be damped with a fine syringe and closed up completely
till 6 p.m. when the light should be opened about an inch for the
night. As the sun gets hot in the morning open it wider ; the
temperature should only vary between a maximum of 85 degrees
and a minimum of 68. Pinch off the shoots at one leaf beyond
the first fruits ; peg the young growths evenly over the soil and cut
off the cucumbers as soon as they are fully formed. Top dressing
or artificial manure watered in will be helpful when the plants have
been bearing for some time. Melons may be treated in a similar
fashion except that the flowers must be fertilised.

When a female flower (it has the beginnings of a melon at its
base) is nicely opened, take a well opened male flower, tear off the
petals to expose the pollen and rub it gently on to the anther of the
female flower. This should be done between 11.0 and 3.0 on a sunny
day when the flowers are dry. Fertilise more flowers than you

require as they will not all set. A melon plant will not often bring more than six or seven melons to perfection.

To make a vegetable marrow bed, collect the mulchings from the rose beds, etc. and put them in a heap and cover them with fine soil. Plant the young marrow plants which may have been raised in the cucumber frame as soon as the temperature has gone down to 75 or 80 degrees, or sow them singly in the heap, covering them with a 4-inch pot at night till they are 2 or 3 inches high, and their second leaves are well formed.

Asparagus may be ready to cut during this month, as soon as it is 5 to 8 inches tall. Cut the stem one or two inches below the soil, taking care not to cut off any young shoots at the same time. The thin pieces can be used for soup, etc.

Indoors. Pot up well rooted cuttings of indoor chrysanthemums and put them out-of-doors at the end of the month if the weather is mild ; if the weather is cold keep them under glass or in a frame. When out-of-doors they should be in full sun and well watered.

To make bushy plants for cut flowers pinch them to 4 inches from their base. Use leaf mould and decayed turf for potting and do not put them out-of-doors immediately after potting unless they have been hardened off already.

Sow tender annuals such as French marigolds, and most other seeds both annual and perennial ; biennials can be left till next month. Prick out celery, tomatoes and any other seedlings into pots or boxes if they are getting crowded or cannot be planted out for some time.

Hints for weekly lessons.

First week. Sow seeds indoors. Outdoors, sow peas, broad beans, carrots, beetroot, etc. and potatoes. Mow and roll. Harden off sweet peas. Prune roses. Plant perennials and biennials. Sow annuals in open border.

Second week. Sow seeds—scarlet runners and green crops. Make the cucumber bed. Harden off seedlings. Prune roses. Plant perennials.

Third week. Sow French beans, peas and greens. Make the

marrow bed. Plant out sweet peas and seedlings. Harden off lobelia, geraniums, etc. Pot chrysanthemum cuttings.

Fourth week. Prick out seedlings. Stick peas. Plant out bulbs. Make the new violet bed. Plant dahlias.

MAY.

Outdoors. The violet bed must be made if that is not already done; sweet peas sticked, the beds kept weeded and hoed, and cleared of spring bedding; polyanthus can be taken up and divided and re-planted in a nursery garden for autumn bedding. Seedlings should be hardened off and pricked out in nice moist weather as soon as possible ; seed patches must be thinned out and the seedlings thrown away or transplanted. When planting seedlings in their permanent places think well of the succession and effect of the colours ; put only strong growing things where they are likely to be over-shadowed, or where a gap has to be filled later ; when planting near a grass border, plant 8 inches inside the bed so that the plants will not be spoilt by the mowing machine. Remember that good sized clumps of things that last a long time in flower are to be preferred to long thin lines and small dabs.

Finish all the planting in the rock garden if possible.

Prick out perennials and biennials in a spare piece of ground. In the middle of the month wallflowers and forget-me-nots should be sown for next year. Old plants of forget-me-nots may be pulled up and put by the heels (*i.e.* roughly planted close together in a small trench and partly leaning towards the ground) in a semi-shady corner to sow themselves ; the seedlings should be taken up later and pricked out when an inch high. Perennials and biennials may still be sown. Poppies and other herbaceous things must be tied up as they require it ; and roses should be looked over, syringed for green fly and the caterpillars picked off.

Cut back cotoneasters while still in flower ; long straggling growth that has not flowered should be cut back almost to the base ; cut off also pieces that grow forward and see that only enough is left to cover the wall space properly.

Cuttings of lavender and shrubby veronicas may be struck out-of-doors in open ground.

Geraniums, marguerites, lobelia and other plants may be planted out.

Ferneries can be made ; plant the ferns under trees or still better in half shade or in an easterly position.   A few large stones may be put to shelter the more delicate ones.   Among those that grow well are lady fern, male fern, hart's tongue, shield fern or polystichum aculeatum, polypody, wall rue and the spleenworts.

Plant out green vegetables as soon as they are ready (*i.e.* about 4 or 6 inches high) both from seed-pans and seed-beds.

Round hoe early potatoes—that is, draw the soil up round the young growth.   Thin out beetroot, carrots and other green crops as they require it.   Celery may be planted at the end of the month or in June ; but the trenches should be dug and well manured as soon as convenient.   Continue sowing peas, French beans, scarlet runners and, in the north, broad beans.   Stick peas and beans. Plant cucumbers, melons and marrows if it was not done in April. Dress the asparagus bed with a slight covering of salt.

Tomatoes may be planted out in favourable weather at the end of the month.

Indoors.   Prick out asters into boxes as soon as they have two leaves besides the seed leaves, and any other crowded or delicate seedlings.

Pot up begonias in a light sandy soil as soon as they have begun to shoot.

Continue to pot up chrysanthemums and put them out-of-doors.

Hints for weekly lessons.

First week.   Tidy the beds.   Plant out seedlings and asters. Sow peas, etc.

Second week.   Plant out geraniums, etc.   Sow seeds of biennials and perennials in the open or in boxes.   Pot chrysanthemum cuttings.   Hoe potatoes.

Third week.   Plant begonias, greens and seedlings of biennials and perennials.   Sow peas, French beans, etc.

Fourth week. Plant out seedlings. Thin root crops. Stick peas. Syringe for green fly if necessary.

JUNE.

Outdoors. In the flower-garden the result of the work during the spring months is beginning to show. Most of the pricking and planting out ought to have been finished but all will have to be kept carefully hoed and weeded. Hoeing lets air into the soil and makes it cool and keeps it ready to suck up any rain that may fall, so hoe occasionally if the ground is becoming hard even if there are no weeds.

Keep everything that requires it well tied up; remember to tie before it is really required : tie loosely and tighten as required, the plants then keep in a good shape. Seedlings often require three stakes at the edge of the clump, round which rafia should be tied lightly. A stake can often be added later to support tall-growing things, such as dahlias and Michaelmas daisies.

In a carnation bed, if full of flowers, the stems can be well supported by sticking in a quantity of twiggy branches about 18 inches to 2 feet high and letting the carnations support themselves on them, such stakes look nicer than wood stakes and hardly shew.

Sweet peas must be looked at occasionally to see that they are clinging to their sticks and the flowers should be picked as soon as they begin to fade or before. If this is done regularly they will go on flowering late into the autumn. If the weather is dry a thorough soaking of water once a week is a great help to them ; manure may be added in one of the artificial forms ; either sprinkle it on or insert it in the ground before rain or watering ; or use it dissolved in the water. If the soil cakes round their stem, hoe it, or lighten it with a hand-fork.

Roses may need syringing with a mildew destroyer, or insecticide for green fly ; follow the instructions which will be printed on the tin ; or you can make the following mixture : beat up a lump of soft-soap, about the size of an egg, in hot water, add a gallon and a half of tepid water and four tablespoons of paraffin oil and mix well. Syringe the roses with this on a dull day or towards evening, if possible using a spray syringe which wastes much less of the liquid : repeat next day if necessary and after that syringe with clear water.

Any suckers from briars on which roses are grafted must be cut off as near the main stem as possible ; their foliage is usually quite different from that of the real rose and their growth is coarser, but if there is uncertainty whether the shoot is briar or rose, grub round the stem below the soil and see whether it springs from above or below the graft ; if from below cut it off. Roses are sometimes grown on their own roots and sometimes grafted on to briar stocks ; if the briar is allowed to grow the grafted rose will be starved and die. Ramblers and Wichurianas are more often grown on their own roots, while the more delicate tea roses are grafted. Dead roses should be picked, snapping them off near the head.

Asters may be helped with liquid manure as soon as they are well established ; they will probably have only been pricked out at the beginning of this month ; they must always be planted in rich soil.

Dahlias and other quick-growing things will appreciate a little help in the way of manure if the ground is really poor, but as a rule they do satisfactorily without it. The same may be said of all annuals, they will do nicely without manure but will be better with it ; the same rule applies to watering in hot weather, but if you do water soak thoroughly once or twice a week, don't sprinkle every day ; if a little rain falls, add to it and turn it into a good shower for the plants.

If things flag from drought or after transplanting they must have regular and good watering for a few days. Take a piece of stick to see if the water has gone down ; you may think you have watered really well and will be surprised to see that often it has only sunk half an inch or an inch down.

Keep the surface hoed and loose so that water and air can get in freely ; during bright weather watering must be done in the early morning or in the evening.

Seeds of biennials and perennials if not already sown should be sown as soon as possible. Self-sown seedlings of forget-me-nots may be transplanted in suitable weather during the summer, and

planted 6 inches apart. Wallflowers which ought to be well up by
now must be pricked out when 4 or 5 inches high into rows 8 inches
apart. It should be done this month or they will not have time to
become good bushy plants ; when established and growing again
nip out the top shoot leaving three or four pairs of leaves below, to
make them break and get bushy.

The last of the bedding and pricking out should be finished as
early as possible ; most of it will have been done in May, but asters,
kochias and begonias will all have been left to this month, and a
few other things that for some reason or other have come on slowly.
If the weather is dry and there is no sign of rain, water the beds well
an hour or so before planting and water again in the evening if
required and for a few evenings afterwards.

All bare gaps from bulbs having died down and spring plants
taken up must be filled or plants trained to grow over them. The
bulb leaves will probably have died down, *i.e.* turned yellow, when
they may be pulled off ; if they come quite easily it is time to take
them off, but if they are not quite ready, twist or plait them into a
knot or tie them together if they are in a mixed border so as to take
as little room as possible. Thin out clumps of annuals and put the
seedlings into the bare places ; it is practically useless to attempt to
transplant some seedlings, though occasionally they will succeed,
such as poppies, eschscholtzia and other tap-rooted plants (*i.e.* with
carrot-like roots).

Carnations should be disbudded if they have too many buds on
one stem ; snip off buds clustering round the top centre bud and
thin those down the stem.

Early flowering shrubs should be pruned by cutting off this
year's flowering stems so as to encourage the buds at their base to
send out young wood for next year.

Any violet runners from the young plants should be pulled off ;
and the violets well syringed with water and, if necessary, with
sulphur if red spider is shewing ; its signs are sickliness and a brown
spotted appearance on the leaves. It usually only comes during
or after drought.

Fruit should be netted as soon as it begins to shew colour ; black currants however are rarely touched by birds and can usually be left unnetted if they are to be used as soon as they are ripe.

As soon as the fruit is set, mulch strawberries with straw to keep the fruit off the soil unless the autumn dressing has been left on, which will be washed clean by this time. All faulty or mildewed fruit should be picked off and thrown away when gathering the ripe fruit.

Stop cutting asparagus in the middle of this month, or a week later in the north : if it is cut longer it does not get its crowns well ripened and the bed becomes too much exhausted to throw up good shoots next year ; asparagus beds if treated well last 40 or 50 years.

The last sowing of peas may be made ; sown later than June they will not pod satisfactorily. Other peas must be sticked as they require it. Celery must be planted in trenches 9 inches below the level of the ground in well manured soil and watered till thoroughly established.

Lettuces may still be sown and can be grown with advantage on the ridges between the celery trenches; they will have been cut before the celery has to be earthed up.

Dig up early potatoes : sow French beans in the first week for succession, also main crop turnips, spinach and radishes. Thin out beetroot, carrots and onions ; the first 8 inches apart, the others 6 inches.

Plant out all kinds of winter greens and cauliflowers 1½ to 2 feet apart each way.

Nip off the top shoot of broad beans before or as soon as the black fly appears, that is, as a rule, when the beans are 3 to 4 feet high.

Keep all the garden hoed and weeded.

Indoors. A few late perennial seedlings may be pricked out into boxes to be planted later on.

Chrysanthemums in pots should have their final potting towards the end of the month and be staked ; for ties, the fasteners in the form of cut wire rings, are an enormous boon ; they save a great deal of time and can be shifted as required ; the cost is practically nothing as they can be used again and again if taken off when the plants are cut down. They are also most useful for carnations.

The chrysanthemums should be planted into 8-inch pots, or two plants in a 9-inch pot and kept well watered ; and from a fortnight after potting they should be fed once or twice a week with some liquid manure or soot water or other stimulant. Syringe for green fly if necessary with a similar mixture as for roses, or sponge the tips of the shoots gently with the soap mixture. Nip out the top shoot again when the stems are 18 inches high if bushy plants are required.

Dirty or empty pots should be washed and greenhouses scrubbed out. All except very delicate plants may be left out-of-doors safely for a few nights while the woodwork is being washed and painted.

Hints for weekly lessons.

First week. Tie up plants. Plant out delicate things. Stick sweet peas. Sow French beans. Mulch strawberries if not yet done.

Second week. Hoe and weed. Finish planting. Thin out seedlings. Plant celery. Net fruit.

Third week. Hoe and weed. Prick out wallflowers and forget-me-nots. Plant winter greens.

Fourth week. Thin out seedlings, beetroot and carrots, etc. Earth up potatoes. Disbud carnations. Pot chrysanthemums.

JULY.

Outdoors. Perennials sown in June will want pricking out into a nursery bed ; it is better as a rule not to put them in their permanent place as the luxuriant growth of older plants overpowers them.

Early annuals may have to be cleared away and strong seedlings which have been kept back for the purpose should be planted in their places ; but with careful planning strong growing later annuals such as the marigold class ought to fill up the gap if they have been planted close by.

Weeds must be kept carefully under, asters fed, and sweet peas trained to their sticks.

If a new stock of iris is wanted, old plants should be divided up any time after flowering and planted in a spare piece of ground : bits may also be taken off Michaelmas daisies and other similar perennials and grown on, ready to be planted out in the autumn.

Delphiniums should be cut down at once when they have finished flowering, then in some seasons they will flower again in the autumn.

Old plants of eschscholtzia should also be cut down; bulbous and tuberous rooted things should be allowed to die down, fibrous rooted things should be cut down as soon as they have finished flowering. In all cases remove dead flowers unless seed is required.

Bulbs which have been lifted and dried off may be replanted or else packed away to plant in the autumn.

Nip down wallflowers to make them bushy—clear away runners from the young violet plants.

Dahlias will want staking by the end of the month.

Deutzias, ribes and other flowering shrubs must be pruned after flowering : hedges will want clipping.

Carnations must be layered by half cutting through the stem of a young shoot just below a bud, in a slanting direction ; then peg the layer into the ground keeping the cut open and cover the stem with a little mound of light soil ; hairpins or a forked bit of wood can be used as pegs. The layers should be rooted in about a month, when they can be cut off and left where they are for a time.

Outdoor chrysanthemums will want staking, the stakes should be 3 or 4 feet high, and those in pots must be kept free from green fly and disbudded if showing flower buds.

Straw can be cleared from strawberry beds when the plants have finished fruiting and layers taken. These can be taken either (1) by cutting them off and planting them in thumb pots or boxes, keeping them well watered ; (2) sinking thumb pots in the ground and pegging the layers down into them and cutting them off when well rooted ; (3) pegging the layers into the ground, cutting them off when rooted and potting them up. In any case try to have the layers ready to plant out early in August.

Raspberries should have unnecessary canes cut away so that

the fruiting canes for next year may have as much sun to ripen them as possible ; only four to six canes to a plant should be left. If a fresh stock of plants is wanted, leave some canes growing a few inches away from the parent plant and remove them in the autumn.

Celery should be earthed up for the first time ; fill up the trench level with the surrounding ground ; take care not to let the dirt get between the celery stalks. If you have no helper to hold the plants while you fill the trench round them, each plant should be tied round with paper or at any rate with bass before beginning to dig.

Second early potatoes should be lifted ; if there are any signs of potato disease, spray the foliage of all the potatoes with Bordeaux mixture ; burn diseased potatoes and foliage ; do not leave them on the ground.

Cabbage patches can be cleared.

Net gooseberries, and take the netting off the strawberries.

Continue to stick peas and scarlet runners. The latter need tall strong stakes but they need not be branching. Put them in fairly close, about a stake to a plant, and strengthen them by poles tied horizontally on either side of the row about 4 feet from the ground.

Indoors. Finish potting chrysanthemums. Go through the indoor plants and weed the pots ; cut off dead flowers and leaves ; top dress them with fresh rich soil if the pot is looking poor and empty. If green fly appears fumigate the greenhouse. Liquid nicotine vapourised over a small spirit lamp is the most effective form of fumigator.

Pot begonias into larger pots if they are growing fast—also cinerarias. The final pots for cinerarias should be 5-inch or larger.

Sow schizanthus for flowering indoors in the early autumn ; when the seedlings are nearly 2 inches high, pot them separately, and re-pot at intervals till their final potting is one, or at most two, plants in a 7-inch pot. They must be nipped back frequently to make the plants bushy. Feed them when they begin to get large plants and are in their final pots.

Hints for weekly lessons.

First week. Stick peas. Prick out perennials. Hoe and weed. Pot chrysanthemums.

Second week. Take strawberry layers. Make a stock of young iris and other perennials. Gather fruit.

Third week. Earth up celery. Prune flowering shrubs. Take carnation layers. Hoe and weed.

Fourth week. Sow schizanthus. Clear out unnecessary raspberry canes. Plant or pack away dried off bulbs. Stake dahlias, chrysanthemums, etc.

AUGUST.

Outdoors. *Early annuals can be pulled up, any seed that is required from them should have been gathered when it is ripe on a fine day. English and Spanish iris may have their leaves cut off ; do not pull, as the bulbs are very lightly rooted.

*Cut down delphiniums as soon as they have finished flowering, in some soils they will flower again in the autumn.

Keep sweet peas cut, removing all dead flowers and seed-pods at once.

*Sow seed of rock plants gathered fresh from your own plants as many rock plants grow much better when raised from fresh seed, especially some of the primulas ; bits of plants may often be taken off and struck to make fresh plants for the spring.

*Re-plant madonna lilies and iris stylosa.

Some perennials may still be sown but it is getting late for them to make good plants before the winter.

*Wallflowers must be nipped down occasionally to make them grow into good bushy plants.

*Stake all the tall plants that require it.

If a plant looks ill and yellow and the soil round it is black and mossy, the soil is sour. Take the plant up, dig the soil well and drain it by putting in broken pots or hard rubbish and fill the hole with fresh soil mixed with sand and road scrapings.

Note. In schools, work marked * may be done in July.

*Wichuriana and rambler roses will require their young shoots to be tied and trained ready for autumn pruning ; shrubs must be pruned as soon as their flowering is over.

Rose cuttings may be struck ; pull off a piece of rose shoot of this year's growth with a heel of last year's wood ; press it deeply into the ground and do not remove it till either next spring or autumn.

Many shrubs will strike if low-hanging boughs are scratched with a knife and bent down and buried in the soil. Put a big stone on the bough to keep it in its place.

*Keep a good look-out for green fly on chrysanthemums and wash it off as soon as it appears. Pinch back for the last time, if bushy plants for cutting are being grown ; go through the flower buds, thinning them out and only leaving one to three buds on a stem according to the size of the variety and growth of the plant. Do not rub off all the buds or little shoots at one time but rub off one or two at intervals of a few days so that the plants should not receive too great a check. The plants ought to be examined every few days, as if the shoots round a flower bud grow more than $\frac{1}{2}$ or $\frac{3}{4}$ of an inch, the bud in the centre is too much starved and it will never swell even if the shoots are rubbed off ; in this case it is better to leave the shoots to form fresh buds for late flowering. Keep all the plants well staked.

*Plant out broccoli ; sow cauliflowers, late turnips and lettuce ; if it is very hot and dry, shade the seedlings with leafy boughs picked off trees.

Celery should be earthed up ; peas and beans should be cleared away as soon as they are over and their place taken by late greens.

*Second early potatoes must be taken up.

Weeding and hoeing should be done regularly to keep the soil open and moist.

Early plums and peaches will be ripe ; if there are many wasps pick the fruit before it is quite fully ripe ; it will ripen if put under a meat safe in a sunny window or greenhouse.

To destroy wasps the most effectual way is to put a teaspoon of cyanide of potassium, powdered small, some inches inside the

Note. In schools, work marked * may be done in July.

entrance of the nest in the late afternoon when the wasps are going home, and an hour or so afterwards block up the entrance with sods or earth ; the cyanide stifles them. Great care must be taken in using this poison as it is a very dangerous one.

Tar poured down the hole is another good way of destroying the nests of both wasps and ants.

Plant your strawberries in a new bed as soon as they are well rooted ; the best place is where early potatoes have been grown ; water if necessary.

Summer prune fruit trees and bushes by cutting back the young growth to 6 or 8 inches ; some people simply snap it and leave it hanging down. Black currants should, however, only have about 4 inches cut off as they flower on the new wood.

Indoors. Pot up Roman hyacinths and freesias ; they must be done in August to succeed really well. The hyacinths should be put in the dark and the pots of the freesias plunged to the rim in soil or ashes out of doors, and kept just moist. When frosts begin at night put them in a window or cold frame.

Hints for weekly lessons.

First week. Sow fresh seed of rock plants. Begin to disbud chrysanthemums. Gather ripe seed. Plant out broccoli and greens.

Second week. Stake tall plants. Summer prune. Cut off dead flowers. Lift early and second early potatoes.

Third week. Pot freesias. Summer prune. Plant strawberries. Disbud chrysanthemums. Earth up celery.

Fourth week. Pot Roman hyacinths. Tie and train roses. Cut off dead flowers. Gather ripe seed.

SEPTEMBER.

Outdoors. Pull up and cut off dead annuals and perennials. Flower seeds may be gathered and hardy sorts sown for spring flowering such as sweet peas, love-in-a-mist, Shirley poppy and cornflower ; they will probably stand the winter and flower well. Carnation layers may be planted if well rooted and lilium candidum

must be moved at once if necessary as it is almost an evergreen
and in some districts begins to grow again about the middle of this
month.

Tropaeolums and hardy cyclamen should be planted and, towards
the end of the month, wallflowers and other spring bedding plants
may be put in if the ground is free.

Hardy seedlings should be pricked out as soon as they are fit.
At the end of the month begin to take cuttings of geraniums and
marguerites ; pot up any old plants that may be needed in boxes
of poor soil and leave them out of doors till October. Salvia patens
and scarlet lobelia, where the latter does not stand the winter, may
be lifted and put in boxes of coal ashes when they have finished
flowering.

If the ground is moist turf can be re-laid. Trim hedges. Take
up and re-pot arum lilies, also cytisus, azaleas, etc., the last do not
require to be re-potted every year.

Go through the indoor chrysanthemums for the final disbudding
of the later kinds ; feed them frequently with soot and manure water,
at this time once a day will not be too much for about a month,
if they look at all starved ; tie up loose stems and rub off side shoots
from the main stem if large flowers are required. Keep the pots
weeded. Bring them under shelter when the buds begin to shew
colour or if it is very cold at night ; syringe them if they are attacked
by fly.

Bulbs may be planted out of doors ; hyacinths and narcissus
an inch below the surface, tulips 4 to 6 inches deep ; crocus, snowdrops, narcissus, scilla nutans (bluebell) and anemones can all be
planted in the spring garden. Woodruff and Solomon's seal can
be planted in shade or semi-shade.

Prepare holes in the grass for planting specimen shrubs ; the
hole should be 2½ feet deep and 3 feet across ; fill in a foot with
rubble and then put in a layer of good soil and manure and leave it
till the planting time in October or November.

Clear out old melon and cucumber plants and fill up the frame
with good soil or from the old marrow bed, and plant violets
6 to 8 inches apart, disturbing their roots as little as possible.

Pull off all the runners and put in the plants quite close to the glass.

Vegetables should be taken up and cleared away as soon as the crop is finished.

Celery must be earthed up, and the last cabbages and broccoli planted. Second early potatoes should be lifted during the first fortnight and the main crop at the end of the month or in October.

Rhubarb may be taken up from now onwards for forcing ; dig it up and leave it out of the ground till touched by the first frost or two and then bring it into the greenhouse or warm dark cellar. If it is put in the greenhouse pack a little soil round it and put it in under the stage which can be made dark by boards or black linen hung round and over the plants.

Strawberries should be planted in fresh ground as early as possible.

Sow winter spinach and lettuce. Lift spring onions ; plant spring cabbages during the first week of September. Gather and preserve scarlet runners in jars, spreading them in alternate layers of beans and salt.

Plant globe artichokes 4 feet apart each way ; cut off the young side sprouts with as much root as possible and plant them singly.

Plant seakale by dividing up the old plants and re-planting the crowns with a good piece of root attached, three in a clump, to be covered by pots or boxes in the spring.

Indoors. Pot bulbs either in light soil in pots or in cocoanut fibre in bowls. Fibre should be mixed if possible with a little crushed oyster-shell and a small quantity of charcoal (about the size of a walnut) put at the bottom of each bowl to keep it sweet. Plant the bulbs leaving their tips above the soil and put them out of doors for a couple of days, watering them well. Then cover the pots with fibre and cover the whole with a good heap of leaves outside or indoors—or they may be put in a dark airy cellar ; in the latter case water them slightly about once a month, and when they have made one or two inches of growth, probably about December, bring them gently to the light. I usually prefer to put those planted in soil under leaves, and those in fibre in a cellar.

Hints for weekly lessons.

First week. Clear away dead annuals and cut down plants. Lift second early potatoes. Sow hardy seeds for the spring. Take cuttings.

Second week. Plant seakale. Earth up celery. Disbud, stake and feed chrysanthemums. Plant carnations.

Third week. Pot bulbs. Plant violets in the frame. Sow winter spinach. Take geranium cuttings, etc.

Fourth week. Plant bulbs out-of-doors. Plant spring plants. Take cuttings.

OCTOBER.

Outdoors. Clear the flower beds as soon as the autumn plants have finished flowering—manure them and fork lightly between the plants, disturbing roots as little as possible. Divide and re-plant old clumps ; plant as many as possible of any freshly bought plants and all hardy bulbs.

Dahlias should be lifted as soon as frost has blackened the foliage, and stored in boxes, covered with sand, in a cool dry place ; if the place is very dry sprinkle them with water every 6 or 8 weeks. Gladiolus and other tender plants should be lifted and kept away from frost. Lobelia cardinalis will stand the winter in some parts of England if the plants are covered with bracken or ashes.

Shorten straggling shoots of roses and thin out weak and unhealthy wood ; begin pruning Wichurianas and ramblers, cutting out all wood that has flowered and tying in this year's growth.

Renew turf on lawns where necessary.

Propagate rhododendrons, aucubas, laurestinus and shrubby spireas by layering them like carnations ; put a good sized stone on the layered branch to keep it firmly in the soil. Lilacs are propagated by planting the young suckers.

Plant hardy ferns—divide montbretias.

Dig and manure the ground 2 feet deep for next year's sweet peas.

Pull pinks, arabis and violas to pieces and re-plant the bits.

Double white rocket (Hesperis matronalis) is very useful and should be divided every year, taking off and re-planting the side pieces and throwing away the centre.

Lilies of the valley may be re-planted ; divide the clumps and re-plant the flowering crowns which should be as thick as a pencil. The bed may then be top-dressed with well-rotted manure or leaf mould mixed with mortar rubble.

Spring cabbage may still be planted. Celery should have its final earthing by the end of the month. Lettuce may be sown for spring use.

Bring tomatoes indoors to ripen. Lift and store beetroot in a cool dry place and towards the end of the month carrots, onions, etc. should be lifted, and the main crop potatoes dug up.

Artichokes (Jerusalem) may be left in the ground.

Old crops should be cleared off and the kitchen garden well dug and manured.

Remove the decayed leaves from seakale and mulch with cinder ashes. Cut down asparagus. Plant horse-radish. Gooseberries and currants may be pruned as soon as the leaves have fallen. Gather nuts, apples and late pears. Cut away any suckers from fruit trees ; lift and root-prune them when necessary, cutting through any main tap-roots. Prepare places for new fruit trees as recommended for shrubs in September. Raspberries, apples and pears may all be planted any time after their leaves have fallen, but usually the first week in November is sufficiently early.

Indoors. Mint and parsley may be lifted and put with a little soil in boxes in the greenhouse for winter use.

Take viola, calceolaria, penstemon, marguerite and geranium cuttings, striking them in a frame or cold house.

Deutzias, dielytras and spiraeas may be potted up and put in a frame till required for the greenhouse.

House chrysanthemums as soon as it gets cold at night and give them the same attention as in September.

Pot any bulbs that have not yet been done.

Hints for weekly lessons.

First week. Plant cabbages. Take cuttings. Gather apples. Take up potatoes.

Second week. Pot and plant bulbs. Earth up celery. Divide perennials.

Third week. Lift crops. Prepare ground for sweet peas. House chrysanthemums.

Fourth week. Prepare ground for planting. Dig and re-plant borders.

NOVEMBER.

Outdoors. The borders must be cleared out, dug and manured if not already done ; all the dead annuals have to be taken up, the perennials cut down, lifted and re-planted when necessary, or else trimmed to a convenient size if they have grown too large ; the bits that are cut off can be used to make fresh plants. All the old borders should have a good dressing of old manure lightly forked into them and should be re-planted every few years. Wallflowers and other spring flowering plants must be put in. Bulbs, that were grown last year in pots and dried off, and fresh bulbs should be planted at once.

Lilies of the valley may still be re-planted.

Wichurianas, ramblers and briars can be pruned ; the tips of other climbing roses should be cut back, the shoots fastened up and all weak and dead wood cut out.

Leaves must be brushed up and put in a heap to decay.

Roll and mow lawns at intervals ; this is also the time to top dress them if they are poor and to get rid of the worms by watering them on a dull warm evening with a teaspoonful of chloride of lime in two gallons of water ; the worms will come to the surface and must be brushed up next day. A mixture of equal parts of finely sifted manure, soot and good mould sprinkled fairly thickly over the surface is a satisfactory dressing. The weeding of lawns should be finished this month.

Cut hardy fuchsias almost down to the ground and in cold climates cover the crowns with ashes.

Any shrubberies that want thinning or clearing should be attended to this month, and brooms, choisya, prunus, pyrus, pernettaya macronata (a little shrub with coloured berries liking damp and shade), and many others may be planted.

Also plant roses ; a very windy garden can be well sheltered with sheep hurdles, or a rose garden can be made inside a small raised mound on the top of which are placed sheep hurdles up which ramblers may be grown. Hurdles can also be put at intervals along herbaceous borders, making a kind of sheep pen and with climbers growing over them these can be made to look quite pretty in a windswept garden.

All the planting of roses, fruit, shrubs and trees should be done this month unless it is very cold and wet. Dig a hole large enough to take the roots nicely spread out and a little deeper than necessary. Fork in some old manure at the bottom ; if fresh manure has to be used, spread a layer of soil over it ; put in the tree or bush and fill up the hole with manure and soil, giving the tree a good shake at intervals to work the soil among the roots. If the soil is wet and very heavy, dry prepared soil must be used to work in round the roots. See that the bush is planted straight and trodden in firmly. Stake it if necessary ; if the manure is hot, use soil only for filling in the hole and put the rest of the manure on the surface as a top-dressing.

A good mulching of manure should be put round the old fruit-trees and roses, or this may be left till the spring and put on just as the young wood is beginning to grow.

Clean the strawberry bed by cutting off all the runners and all dead and dying leaves round the plants and mulch with strawy manure ; this may be left on the bed during next summer for the fruit to rest on as the winter rains will have washed it quite clean. The bed must be well weeded and the ground between the rows lightly forked before the mulching is put on. Do not mulch young plants.

Potatoes and root crops must be lifted, and as much as possible

4—2

of the kitchen garden dug and manured. Slow acting manures such as basic slag should be applied now. The ground should be left rough to allow the frost and air to penetrate it. If it is infested with insect pests a good dressing of gas lime should be given and left on the ground till spring ; as this is a virulent plant poison it must be applied at least three months before the ground is to be cropped.

Globe artichokes should be mulched with strawy manure or ashes to keep out the frost.

Begin pruning fruit-trees and cut out all dead and weak wood. Dig and manure the ground for the first sowing of peas in February.

Indoors. Freesias should be put in a frame or cool greenhouse when the nights get cold, and chrysanthemums put under glass or indoors. Those that have already finished flowering may be cut down and put near the glass to encourage nice sturdy cuttings ; too forward sprouts should be cut off.

Begin to dry off indoor geraniums and fuchsias—they will need practically no water from now to February.

Look at Roman hyacinths and early narcissus and bring them to the light as soon as they have made an inch of growth.

Hints for weekly lessons.

First week. Lift crops. Clear strawberry bed. Prune ramblers.
Second week. Dig and re-plant borders. Plant roses.
Third week. Dig kitchen garden. Plant trees and shrubs.
Fourth week. Attend to early bulbs. Dig. Prune fruit-trees.

DECEMBER.

Outdoors. There is little to be done in the flower garden beyond cutting down and tidying up after the latest flowering things are over.

Rock gardens may be made in open weather and any up-rooting that has to be done should be seen to.

Gravel paths if mossy should be picked up with a small pick-axe and raked and well rolled, keeping them rather high in the centre.

Turf may be re-laid in mild weather from now to March. Lift it evenly and lay it on level raked ground and pack in soil underneath where necessary to make it smooth and even; when laid, batten (or beat) it well down with the flat of a spade and roll well. If roses arrive during frost leave them in their wrappings, but if the frost lasts, hack up a piece of ground and put them in by the heels. They must on no account be planted while they or the ground are frozen.

Outdoor chrysanthemums should be cut down to a few inches from the ground.

Christmas roses may be protected by bell glasses or by a frame placed over them. Set the frame on four bricks to allow the air to circulate freely among the plants.

Rubbish from burnt heaps may be dug into the soil.

Any renovations of pergolas, paths, edges and rain-water tanks should be done this month.

Finish as much of the pruning and digging as possible; the ground must be dug up rough, leaving it in large clods.

Trunks of fruit-trees covered with moss or lichen should be dusted over with slaked lime or washed with a solution of lime having the consistency of cream.

Scrape trees infected with American blight and wash them with a mixture of lime (5 lbs.), sulphur (1 lb.), and water (2 gallons), or paint them with potash and quicklime dissolved in water to the consistency of cream.

Indoors. Lilies may be re-potted and put into a cold frame and the pots packed round with cinders.

Early bulbs covered with leaves or in fibre should be looked at and if they have made an inch or an inch and a half of growth they should be brought gradually into full sunlight.

Mustard and cress may be sown indoors.

Stored apples should be looked through and rotten ones thrown away; if apples are wrapped separately in newspaper and packed in boxes they keep good and plump for a longer time, and the bad ones do not so easily infect others near them.

The indoor chrysanthemums should be cut down when they have finished flowering and if the cuttings at the base of the plants are growing fast, either the pots should be put near the glass or the tallest cuttings should be pulled off and thrown away, so as to encourage a later growth of cuttings. Unless you have plenty of heat it is better to postpone taking cuttings till February.

Seed boxes can be prepared for spring work, labels got ready, potting sheds thoroughly cleaned.

In open weather some soil may be sifted and put under shelter as often the potting heap out-of-doors is too cold and wet in the winter for use.

Hints for weekly lessons.

First week. Prune fruit-trees. Prepare supply of sifted soil. Cut down chrysanthemums.

Second week. Dig. Prune fruit-trees. Prepare seed boxes.

Third week. Cut down chrysanthemums. Prune. Clear shrubberies.

Fourth week. Pot indoor lilies. Dig. Prune. Renovate fruit enclosures.

CHAPTER V

VEGETABLES. ROTATION OF CROPS. LIST OF USEFUL VARIETIES. STICKING PEAS, ETC. POTATOES. GREEN VEGETABLES. ROOT CROPS. PROTECTION FROM BIRDS AND SLUGS. LETTUCE. CELERY. ASPARAGUS. GLOBE AND JERUSALEM ARTICHOKES. MUSTARD AND CRESS. SEAKALE. CARROTS AND BEETROOT. LEEKS. WHEN VEGETABLES SHOULD BE GATHERED. DATES FOR SOWING.

The ground in which vegetables are to be grown should be well dug and manured. Root crops such as carrots prefer a fine sandy stony soil trodden quite firm, while celery and cos lettuces like a similar soil enriched with good manure. Practically speaking, however, most vegetables will grow in any ordinary soil. If the ground is very hard the surface should be well hoed frequently, especially while the plants are small, and should have been well broken up before the seed was sown.

As far as possible root crops and green vegetables should be grown in rotation; that is, where roots were grown one season, green vegetables should be grown the next year. Early potatoes for instance are an excellent crop to grow in the ground where a new strawberry bed is to be planted in the autumn.

Though it is often advisable to ask some local gardener which varieties grow best in the neighbourhood, the following are a few names of some good kinds.

Peas. Dawn. Rent Payer. Commander. The Gladstone. Sugar Pea.

Broad Bean. Early Longpod.

French Bean. Canadian Wonder. Golden Butter.

Runner Bean. Champion Scarlet.

Beetroot. Crimson Ball.

Kale. Tall and Dwarf Green Curled.

Broccoli. Self-protecting Autumn. Adam's Early White. Purple Sprouting. Late Queen.

Cabbage. Enfield Market. Dwarf Green Curled (Savoy).

Carrot. Early Nantes. Intermediate Scarlet.

Cauliflower. Autumn Giant.

Celery. Giant White. Standard Bearer.

Lettuce. All the year round. Tom Thumb. Paris White.

Onion. Ailsa Craig. Brown Globe. White Spanish.

Parsley. Champion Moss Curled.

Parsnip. Hollow-Crowned.

Potato. Express. Ringleader. Snowdrop. Up-to-Date. Factor.

Radish. French Breakfast.

Spinach. Longstanding Round or Summer.

Tomato. Early Prolific. Golden Drop.

Turnips. Golden Ball or Orange Jelly.

Vegetable Marrow. Long White. Pen-y-byd.

Peas must be sown in well enriched ground. Dig a trench with a spade or mattock 6 inches deep and 6 wide. Scatter the peas an inch or two apart over the bottom of the trench and cover them with 4 inches of soil. When they have grown 3 or 4 inches draw up the rest of the soil round their stems. Stick them when they begin to grow.

French beans are sown in a similar trench, but the seeds are put in a double row 4 inches apart. *Broad beans* are sown 6 inches apart each way, and *Scarlet runners* may be put in a single or double row 8 inches apart.

If birds are troublesome cover the young peas with peaguards or strain black cotton along the row ; if they are very persistent strain the cotton at varying heights till it is 15 inches or more from the ground. It will also serve to support the peas till they are properly sticked ; it is of course not removed till the whole row is cleared away. The seeds must be dipped in paraffin and dusted over with red lead before sowing if the field mice eat them.

The sticks for peas should be twiggy right down to the ground ; they can be used for two years, but after that they become rotten and should be used up as firewood ; they must be from 4 to 6 feet high according to the variety of pea grown.   For scarlet runners the sticks need be little more than thin poles, but should be about 8 feet high.   Plenty of sticks must be used, stinting them only results in the row being spoilt by rain or wind.   The poles for scarlet runners should have horizontal poles, placed on either side of the row about 4 feet from the ground, and tied together, holding the row of sticks firmly between them.

As a rule seed should not be kept from year to year ; if it is more than a year old it cannot be relied on to grow and in any case germinates much more slowly than new seed.

The first sowing of green vegetables may be made in boxes in February and March, to be hardened off and pricked out later on : to get very early peas and French beans the seed must be sown under glass in January or February and pricked out in April or May ;   or the French beans may be potted into large pots and cropped under glass.

It is useless to sow peas later than July as they never pod properly.

Main crop potatoes must not be sown in the same ground more than two years running ;   the ground they enjoy most is that which has been well enriched, as for celery, the year before, and only lightly manured in the previous autumn.

The seed potato with one or two eyes just sprouting should be planted 4 inches deep in April—the earth is then usually left in low furrows along the rows.   After the potatoes have made a few inches of growth, hoe the whole bed carefully, and if the ground is full of weeds this must be repeated once or twice ;   by this time the ground will have been levelled by the hoe.   When the potatoes are about 8 inches high draw up the soil round their stems, making the ridge about 9 inches high from bottom to top ;   this is best done with a mattock or bent fork ;   these final ridges cover the tubers with a good depth of soil and prevent their pushing their way to the surface.

When the crop is being lifted, all diseased potatoes must be carefully picked out and burnt, or the germs of the disease will be absorbed by the soil to break out next year during wet weather ;

the signs of disease are dark, blackish, unhealthy-looking spots which soon develop into rottenness.

If the crop has been badly attacked by disease the stored potatoes should be looked over again in about a month's time and the diseased ones picked out. Disease is much worse in a wet season. If the potato patch is small the foliage may be syringed with Bordeaux mixture, but the labour without a proper syringe is too great if the crop is large; Bordeaux mixture being a strong poison, must be used with care.

Seed potatoes should be sorted out for next year's sowing—they should be rather small, about 2 or 3 inches long, and if any have become at all green by being exposed to the air they should be put among the seed as they are no longer good to eat. In the spring the seed potatoes should be laid out on shelves or in boxes to ripen in full daylight and have all their long sprouts rubbed off; the sprouts when planted should not be more than half an inch long.

All other potatoes must be kept in the dark or covered up.

Green vegetables such as broccoli, sprouts, etc. are sown in boxes indoors or in drills in the open; they are then pricked out once or twice and in their final places they must be allowed plenty of room as they are greedy growers and exhaust the soil quickly; 18 inches apart one way and 2 feet the other is the least space to allow them, except savoy cabbages which may be a little closer.

Root crops such as onions, beetroot, carrots and salsafy should be sown out-of-doors in rows 15 to 18 inches apart and the seedlings must be thinned out later to 4 to 6 inches; beetroot should be allowed a little more room, about 8 inches, while radishes need only just be thinned an inch apart in a single row. Sow the seed evenly and thinly.

These seeds often need protection from birds or slugs; against the former use peaguards or black cotton strained along the rows a few inches above the ground. Put in a few short sticks at either end of the row and strain the cotton from one end to the other. Soot, soot and lime, or alphol are all good preventives against slugs; scatter the two former round the seedlings; alphol may be scattered over the seedlings, as unlike soot it does not burn the foliage.

Lettuces may be sown both indoors, and outside, and then pricked out. Cos lettuces have to be tied with bass as soon as they begin to heart ; this stage is shewn by the leaves beginning to curl over the centre ; tie them lightly about two-thirds way up the leaves. They grow exceedingly well pricked out on the top of the ridges between the celery trenches and are ready to cut before the celery has to be earthed up.

Lettuces must always be pricked out before they get at all drawn and lanky, and should be cut as soon as they have well hearted. They should never be sown thickly as it is apt to prevent them hearting when pricked out. Winter and early spring lettuces should be planted out in the autumn, if left till the spring they are inclined to bolt or run to seed.

Celery is grown in trenches about 8 inches deep : the soil at the bottom of the trench must be well dug and manured a month before the celery is planted. First sow the celery under glass, then prick it out into boxes and keep it under glass till it is nearly 5 inches high, when it may be put out-of-doors to harden. After it has made a little more growth plant it in the trenches 10 inches apart in a single or double row ; if in a double row, plant the second row chequer-wise opposite the centre of the gaps between the plants in the first row.

The plants must be well watered in dry weather till they are established. When they are about 15 inches high, usually the beginning of August, pull in some of the earth round them with a trowel. In a short time put in more earth almost filling the trench. Some one else must hold the stalks together while the earthing up is done, or the plants must be tied up with bass to prevent the soil getting inside the plants and causing decay and brown spots to appear. When 6 inches more growth has been made they must be earthed up again to just above the paler leaves in the centre of the heart, and at the end of September they must have their final earthing up which should make the celery ridges 18 inches to 2 feet high from the bottom of the trench to the top of the ridge.

Asparagus beds must have a good mulching of manure in the autumn ; about March this is raked off and salt is scattered over the beds (giving the appearance of a sharp hailstorm) ; this is repeated either in about 3 weeks time or half-way through the cutting season, using about half as much salt as for the first dressing. On cold soil it should not be applied till half-way through April. Asparagus must not be cut later than the middle of June in the south and the third week of June in the north, or the bed will weaken and the asparagus get poor and thin.   In the autumn the growth may be cut down as soon as it begins to turn brown.

When planting a bed buy good three-year old plants.   The ground must be dug out to the depth of at least 2 feet in the autumn —loosen the bottom, mix in some well-rotted manure, old bones, salt and lime rubbish—these should fill about a foot of the trench when trodden in.   The next 6 inches should be fibrous loam and on the top of that the old soil mixed with old stable manure, till the bed is 6 inches above the surrounding ground ; if it sinks much during the winter add some more manure and soil.   Plant early in March, cover the plants with 2 inches of rich soil, and after the first rain add another 2 inches.   Asparagus must not be cut the first season and as little as possible the second season.

Globe artichokes are increased by having the young crowns or clusters of leaves broken away, with some root if possible, from the parent plant, either in spring or late summer ; they do much better if they are given a mulching of straw or ashes to preserve them from frost during the winter.

Jerusalem artichokes are simply planted in the ground 15 inches apart and 4 inches deep ; they can be bought from a greengrocer, or saved from last year's stock, and planted in the spring.   They will do well in a shady corner where other things will not thrive.

Mustard and cress are best sown under glass as they are less likely to get gritty—when they are 2 inches high they are ready for cutting.   Sow the seed in a box and press it firmly down on to the soil with a brick or something flat and do not cover the seeds with soil.

Plants of seakale may be taken up and divided as soon as they have died down, if the ground is moist ; the crowns are cut apart and re-planted in groups of three, 7 inches apart each way ; straw, bracken or ashes are put over the crowns during the winter and in the early spring, from January onwards, they are covered with pots or boxes heaped over with litter or manure, until it is ready to cut. Seakale can also be divided in February or early autumn.

Horseradish may be grown by any old crown being planted after the root has been used.

Carrots and *beetroot* must be dug up in the autumn and packed away in boxes or heaps covered with a little sand or sacking, secure from frost. When lifting beetroot be very careful not to prick the roots for it will spoil their colour and make them pale.

Salsafy and *parsnips* are hardier, and in the south of England, together with Jerusalem artichokes, can remain in the ground through the winter.

Cauliflowers and *broccoli* need frequent attention as at some seasons they grow very quickly and soon break and spoil, though the broken heads may be cut and are quite good to eat.

Sprouting broccoli grows in tiny shoots and if gathered regularly sprouts again and again and remains a long time in season.

Cabbages should be cut when they feel firm ; the sprouts that shoot out later from them and from Brussels sprouts and other kales can be picked and served as greens or passed through a sieve as a substitute for spinach. Turnip tops can be cooked in the same way. Pick all greens before the flowers open.

Spinach should be cut just before it flowers ; cut it down to 6 inches from the ground and it will usually sprout again for one more gathering ; it can be sown in spring and again in September or October for winter use. Sow it in drills about 2 feet apart.

Leeks should be sown in April and pricked out in June or July, 6 inches to a foot apart.

The following list gives the time for sowing vegetables ; in the north they should be sown a little later.

Potatoes	March	Leeks	April
Broad beans	Feb. to April	Lettuce	(Under glass) Feb. and March
French beans	May to July		(Outside) all the summer
Runner beans	May	Sprouts	March and April
Beetroot	May	Kale	March and April

Broccoli	March and April
Cabbage	April and May, June to August
Cauliflower	(In heat) Feb. and March / (Outside) April
Peas	Feb. to June
Carrots	April and May
Spinach	March and at intervals / End of Sept. for winter use
Jerusalem artichoke	March and April
Vegetable marrow	May
Salsafy	May
Scorzonera	May

CHAPTER VI

FRUIT TREES. PLANTING. PRUNING. PLUMS. PEARS. APPLES. CURRANTS. PEACHES. GOOSEBERRIES. BLACK CURRANTS. RASPBERRIES. THINNING FRUIT. MULCHING. SUMMER PRUNING. STRAWBERRIES.

No exact rule can be given for the depth at which fruit trees ought to be planted but it is generally safe to plant them at the same depth as they have been planted in the nursery garden—the mark on the stem is a guide ; their topmost small roots should be 2 inches at least below the level of the soil.

Make the hole in which they are to be planted large enough to allow the roots to be well spread out without being crumpled. If the situation is exposed to high winds lay the best roots towards the quarter from which the prevalent wind blows, to give the tree good support.

If the tree is being planted against a wall see that the branches are arranged to fit as closely to the wall as possible ; the roots should however be placed at some little distance from the wall, about 6 to 12 inches. Dig the hole a little deeper than necessary, then put in some manure and fork it well in at the bottom of the hole ; cover this with a thin layer of soil ; put in the bush or tree and fill up the remaining space with old manure and friable soil, and a little lime for stone fruit if the soil is at all deficient in it. Fresh and hot manure must not come in contact with small fibrous roots as it will burn them. If fresh manure has to be used put it in the bottom of the hole and cover it with soil ; then plant the tree and fill up with earth only, using the rest of the manure as a mulch on the top.

Trees must be staked if they are exposed to much wind, and wall fruit should be tied to wires or nailed to the wall. Espaliers must have their branches tied horizontally to one or two stakes on either

side.   Three or four branches are sufficient on each side.   Cordon
bushes, which consist of a single stem fruiting the whole of its
length, must be fastened to a wall or tied to stakes or wires ; they
are the most satisfactory, above all for gooseberries, in a small
garden.

Fruit trees should only be pruned lightly after transplanting ;
cut back the young shoots a third of their length except the leading
shoots which should only be tipped back 2 or 3 inches.   When
possible fruit trees should be sheltered from cold north and east
winds as these are apt to spoil the setting of the blossom.

Plums, pears, apples, red and white currants, whether grown as
bush, espalier or wall-fruit must have their yearly new growth cut
back to 2 or 3 inches, leaving only leading shoots longer if the tree
is to increase in size ; these should be left about 9 inches long.

These varieties fruit on two-year old wood and the young shoots
are cut back to throw all the strength of the tree into forming fruit
blossoms for the next year.   Net currants as soon as they swell
and shew colour.

Peaches fruit on the new wood ; the growths should be thinned
out till they are about 8 inches apart and then shortened back
to 6 or 8 inches in February, always cutting the shoot back to
above a leaf bud and not above a fruit bud only, as if there is no
leaf beyond the fruit the sap is not drawn up to nourish it properly ;
the leaf bud may be recognised by being much thinner and narrower
than a flower bud ; it grows either alone or between two flower buds.

Peaches must be planted against a south or south-west wall
and to do well should have a glass shelter projecting about 2 feet
from the wall, over them.   As soon as the blossom begins to open
nets or curtains should be hung over the trees and left till the fruit
is fully set, or drawn down every night and cold day.   The bottom
of the curtain should be tied to stakes or hurdles a yard or more
from the wall to prevent the wind blowing it on to the blossom and
damaging it.

Gooseberries fruit on both old and new wood, but as they are apt
to get quickly overgrown, it is advisable to cut back the new wood
to 5 or 6 inches and to thin out both old and new wood from the

centre of the bush and elsewhere ; if the young wood is left 10 or 12 inches long the old wood must be cut away in proportion. When the bushes get over thick the fruit dwindles and is very difficult to gather. Where cordon bushes are grown, nip back the side shoots to 4 inches in the middle of July to help the fruit to swell and ripen.

Black currants fruit on the new wood. Cut away all wood from which poor shoots are growing or cut it back to just above a strong young shoot. Tip back the young wood 4 or 6 inches and thin out where necessary. The wood should be encouraged to grow from as near the base of the bush as possible, as this allows strong fruit-bearing boughs to be kept at almost their full length without the bush getting straggly.

Raspberries and other canes are grown in rows. The canes should be tied to two rows of wire strained between posts. The first pruning after planting should be severe and the canes must be cut down to a foot from the ground to encourage young canes to grow strong for the next season's fruiting. Every autumn cut away all the old wood completely and tie in four to six canes for each plant ; the first year only one to three should be kept. The plants should be 2 ft. 6 ins. to 3 feet apart and the rows 5 feet apart. Young canes shooting up at a little distance from the parent plant may be dug up and re-planted if new canes are required, otherwise cut off in July all unnecessary canes which will not be wanted for next year's fruiting.

In February tip back the canes to 8 or 10 inches above the top wire ; 6 inches in any case should be cut off. Pruning in the autumn often results in the canes dying back some inches after frost. Autumn flowering raspberries have their old canes shortened to a foot in April to encourage the young growth which bears fruit in the autumn.

During the summer the ground between the rows should be kept hoed, and the old fruiting canes cut away as soon as the fruiting season is over.

Plums, pears and apples should have their fruit thinned if it is growing in clusters and fine fruit is desired. Never leave more than two pears or apples together.

When training wall fruit care must be taken not to let it get overcrowded. The young wood that is nailed in should be at least 6 inches apart and all wood that crosses or rubs another bough, or shoots straight away from the wall should be cut out. The wood requires thinning in bush trees also and all dead and sickly wood cut away. Espaliers are pruned like wall fruit, but only three or four main boughs are allowed to grow on each side.

Mulching with manure may be done in autumn or spring, but the latter often gives a better result. The stimulant helps the trees just as they are beginning to grow and the manure is left on the ground to keep the roots cool in dry weather.

Summer pruning is very important when the growth is vigorous and should be done in July or August ; the sun has then a full chance of ripening the wood for next year. In summer pruning, fruit trees that bear on the old wood should have their shoots cut back to 6 or 8 inches from the base, or the shoot may be snapped but left hanging on the tree, while those that bear on the new wood should have their shoots tipped 4 to 8 inches back.

If fruit trees grow very rank and bear little or no fruit, dig all round and underneath their roots, cutting off any large coarse ones, and especially any tap-root which is running straight down, as this soon reaches the poor subsoil and results in a lack of fruit buds.

Strawberries, which are propagated by runners, should have the runners rooted and cut off as soon as possible and planted in well manured ground in August or September ; in dry weather they must be well watered for some days ; new runners must never be planted in ground which grew strawberries in the previous season. They follow best a root-crop, such as early potatoes. Some new plants ought to be made every year as the whole bed should be re-planted every three years or else the fruit deteriorates.

As soon as the runners are taken and the fruiting is over, trim the old plants by cutting off all the remaining runners and all dead and dying leaves, and hoe and clean the ground well. They may need to be trimmed and hoed again later in the autumn before being mulched for the winter.

In the spring clear off the mulching and hoe the ground before spreading fresh straw on which the fruit will rest to keep it clean ; this should be done as soon as the blossom begins to show, or the mulching may be left, having been washed clean by the winter rains, but the bed must be thoroughly weeded.

If the straw is not put down before the blossom is out, it is better to delay doing it till the fruit is set or the blossom may be damaged. The fruit must be netted as soon as it swells and begins to colour.

NOTES

CHAPTER VII

ROSES. SITUATION. SOIL. PLANTING. PRUNING.
GREEN FLY. LIST OF ROSES.

Roses should be planted in an open sunny situation sheltered from the north and east as cold winds or draughts are fatal.

It is a good plan to prepare the bed some weeks before planting ; the ground should be dug at least 2 feet deep and farmyard or stable manure and basic slag added, and some clay if possible in sandy districts. In very light soil cow manure and chopped turf help to retain the moisture and richness.

Make the hole rather larger than the roots and spread them out before beginning to fill in the soil, but first cut off cleanly any bruised or ragged roots. Plant the rose so that the graft where it joins the stock—a knob-like swelling—is buried an inch below the surface. Fill the hole with dry soil that will work in among the roots, put another small layer of manure about 4 inches below the level of the ground, then fill up with soil. Tread it all firmly, and level the ground with a fork.

If roses arrive during frosty weather leave them in their packing or else put them in by the heels till a thaw comes. Never plant during frost.

Roses may be mulched in November with a good layer of strawy manure, or you may fork it lightly into the bed. The mulch can be cleared away or forked in at the beginning of March.

Pruning. Cut new rose bushes back to five or six eyes from the ground ; new climbing roses must have their shoots shortened a third of their length.

Ramblers have the old flowering wood cut out and the young shoots thinned if necessary and tied in. Old ramblers and Wichuri-anas require a great amount of thinning as every eye sprouts and soon the bush becomes too thick unless it receives attention ; five or six stems, unless there is a great space to cover, is usually sufficient to leave.

Scotch and Penzance briars only require weak wood to be cut out ; in every other way leave them severely alone.

Tea and noisette roses should be pruned at the end of March or in April : if they are pruned earlier there is a risk of the young shoots being damaged by late frosts.

Rambler and Wichuriana roses should be pruned in the autumn but may be left till the spring.

When pruning old bush roses keep as much as possible of the new wood sprouting from near the base, and cut it back to four or six eyes, unless a large bush is wanted when about a foot of growth may be left. If you have to keep new wood sprouting from last year's growth cut it back to four or six eyes ; a very little practice will tell you which is old and which is young growth ; as a rule the young growth is greener or redder and looks more sappy and soft and less woody than last year's growth.

Cut away all thin and sickly wood if there are plenty of young strong shoots, and always try to prune back to an eye pointing outwards as the bush ought to send up its main stem in the direction of the top eye, and would become overcrowded if several stems grew towards the middle. Cut back each shoot to one-third of an inch above an eye ; if more is left it is apt to die back, if less the bud may be damaged.

If roses are attacked by green fly they must be syringed two or three times on dull evenings with a solution of paraffin, soft soap and water, or with an insecticide. If caterpillars appear they must be picked off by hand, or the leaf in which they are sometimes embedded nipped hard to squeeze the caterpillar inside, or picked off and burnt.

The following are some good and useful roses.

Bush. Caroline Testout. Lyon Rose. Joseph Hill. Mrs J. W. Grant. Betty. Killarney. Le Progrès. Mme. Abel Chatenay. Hugh Dickson. General MacArthur. Frau Karl Druschki. Mme. Ravary.

Climbing. Gloire de Dijon. Frau Karl Druschki. Ard's Pillar. Lady Waterlow. Leuchstein. Mrs Orpen. Pink Pearl. William Allen Richardson. Bar le Duc.

Wichurianas, Ramblers and Briars. Dorothy Perkins. Lady Gay. Lady Godiva. Dorothy Dennison. Alberic Barbier. Desire Bergerac. Hiawatha. Jersey Beauty. Paul Transon. Blush Rambler. Crimson Rambler. Lady Penzance. Edith Bellenden. Jeannie Deans.

NOTES

CHAPTER VIII

FLOWERING SHRUBS. PLANTING. PRUNING. MOCK ORANGE. RHODO-
DENDRONS AND AZALEAS. SPIREAS. BROOMS AND GORSE.
AYRSHIRE ROSES. CLEMATIS. CLIMBING SHRUBS. EVERGREENS.

Shrubs should have the ground prepared, and be planted as recommended for fruit trees ; the distance between shrubs varies according to the type of plant, but they must not be overcrowded. Most shrubs require very little attention ; what pruning they need, if any, is simply to cut back the old wood after flowering. Cut out the old wood and leave the young wood to flower next year ; this is especially important for those varieties that make long luscious growth each year such as deutzias and ribes. Lilacs, when small and weakly, should be thinned out if necessary, and their flower trusses picked off as soon as the flowering season is over. As a rule do whatever pruning is required immediately they have finished flowering ; among the exceptions is the purple buddleia which is cut back in the spring, about a third of its length being cut off the young wood and the old wood being cut out ; in some places it is a most vigorous grower and has to be cut back severely.

Those shrubs that make a quantity of twiggy growth require to have it cut back two-thirds of its length and this treatment is given to all that flower on the present season's wood. Big shrubs in a wild garden can usually be left untouched except for cutting out dead wood and trimming straggly branches back to a young shoot or bud near the base of the branch.

The following are some pretty and easily grown shrubs :

Mock Orange or Philadelphus. Very sweet, especially the small flowered variety ; the large flowered and double varieties are very beautiful : var. grandiflorus, deutziaflorus, candelabra plenus.

Rhododendrons and azaleas. These do well in towns and smoky districts; they loathe lime but love peat, and the former do well in semi-shade as well as in the open. If they are weakly they should be mulched with half decayed cow manure. The dead flowers should be picked off. There are numbers of beautiful white, pink and crimson rhododendrons; those with a magenta shade should be avoided. Azalea mollis is the best to grow on the whole and may be had in a variety of shades of pink, yellow and orange.

Shrubby spireas grow well in smoky atmospheres, only, if the air is too dirty, they look shabby and draggled. They grow 8 to 10 feet high and bear long sprays of fluffy white flowers; there are many very pretty varieties and also a dwarf form, S. arguta—3 feet.

Brooms and cytisus are charming. There are the common yellow; the large flowered yellow; brown and yellow (Andreanus); white and cream (praecox); sulphur (sulphureus); rose purple (purpureus); all easily grown.

Double gorse makes a glorious blaze of yellow in sunshine and is well worth its place in the shrubbery.

The white double-flowering cherry, prunus pissardi, single and double deutzia, white and pink weigelia, Judas tree, white and purple lilacs, barbaris Darwinii, stenophylla and ribes are all beautiful flowering shrubs that are well worth growing.

Ayrshire Roses, Bennett's seedling and Félicité Perpétué all do well on a north wall, so does clematis Jackmanni (purple) and C. Anderson Henryii (white).

Among climbers the following are charming. Ceanothus (Gloire de Versailles), pyracanthus, cydonia or pyrus japonica of various shades, yellow jasmine, both winter and summer flowering, white jasmine, and the more tender solanum jasminoides, clematis montana, the white is very sturdy, the pink (rubra) is more delicate; these require their weak and superfluous shoots to be removed—they flower on last year's growth; C. Jackmanni, purple and mauve varieties, which has its growth cut well back to the old wood in the spring; some gardeners advocate cutting down the old wood to 18 inches or a foot from the ground in spring; all poor wood should be thinned out; C. Viticella also has its weak growth removed and the stems which have borne flowers shortened back severely. The honeysuckles (lonicera) are also most useful; those bearing reddish flowers are most showy; cut out old and weak wood and shorten back straggling sprays; they flower on the old and new wood.

Many of these climbers are most satisfactorily grown by being tied to wires fastened to the wall about a foot apart, strained either perpendicularly or horizontally; the former for creepers, the latter for shrubs. Clematis perhaps looks its best climbing over rustic work or a small dead tree, which retains its branches, set firmly in the ground. C. montana rubra is particularly effective trained in this way.

Aethionema grandiflorum is a pretty sub-shrub with masses of pink racemes growing about one foot high and of a spreading habit; it likes a well drained sandy loam.

The garden ought also to contain some of the trees and shrubs that turn a beautiful colour in the autumn such as American oak, rhus or sumach, Siberian crab, maples and others.

There are also many varieties of roses that deserve a place in the shrubbery, such as varieties of rosa rugosa, polyantha, Scotch briar, and single roses.

Evergreens and hedges should be pruned and trimmed in November: hedges also need to be clipped in the summer about the end of June, if a stiff neat effect is desired.

NOTES

CHAPTER IX

PREPARATION FOR THE ROCK GARDEN. TYPE OF ROCKERY. STONE. LAYING IT OUT. PATHS. SOIL. SHRUBS. PLANTING. SEEDS. WATER PLANTS.

If it is possible when you intend making a rock garden in the autumn, make up your mind in the spring and begin to lay in a store of plants. A good many things can be raised from seed and will make large plants in the following year; among these are lychnis, dianthus, rock-rose, forget-me-not, linaria, saponaria and aubrietia.

Next buy or beg any nice rock plants you can get hold of, especially those that will break up into smaller plants; break these up into little bits and plant them in a nursery garden and by the autumn they will have grown into nice-sized clumps, and some may even be divided again. Also take cuttings of double arabis, cistus, veronica, and any other shrubby plants you can, and grow them into really nice plants. Aubrietia and single arabis are best from seed; there are several varieties of the former that are well worth growing.

Sedums, saxifrages (mossy and rosette), campanulas, cerastiums and violas are among those that divide well and increase.

Rock seeds are best sown in boxes in March and pricked out into a nursery bed as soon as the seedlings can be handled. Take care to keep them safe from slugs—dust them with lime and soot or alphol.

When autumn comes start to make the rock garden.

The first thing to decide is whether it is to be a rockery, a rock garden or wall garden. For the last the stone may be of a more uniform size than for the two former; a rough wall has to be built with no mortar but with a small layer of earth between the stones, and the stones must be laid with thought in view of future planting.

Leave nice gaps occasionally between the stones of three or four inches ; if you have any really awkward rooted thing that requires a long root run, plant it as you build. Make some little pockets in the wall holding a good depth of soil which cannot be washed out by rain.

The wall should face south-west if possible, but at any rate it must face partly south. At the base of the wall a stony slope might be formed of soil and of the larger rocks in which to plant things which do not like the excessive dryness of the wall. If the wall supports a bank of earth, these can also be planted near the top of the wall and allowed to grow down it.

Then as to stone. Up to a certain point you must get what is available locally—limestone and sandstone are best but avoid chalk. If there is a quarry where you can get rough weathered stones that have been thrown aside as unsuitable to be squared, get them ; but even if new, let them be rough and not squared. Watering with liquid manure soon improves the colour of new stone. Choose a few large stones, a good number medium sized—not many small ; a ton seems to go no way at all. To make my own rock garden, about 18 yards by 12, I used about 12 tons and I do not think that it could have been made with less. If stone is not obtainable, builders cement mixed and covered with gravel or shingle makes a better substitute than tree roots, old bricks or clinkers.

If a rockery only is the object, don't think the worst spot in the garden under dank trees will do—it won't. Those parts must be used for ferns and for things that don't mind damp shade. Next, don't think you can make a mound of earth and dab down a few large stones and make them firm with a trowel—you can't. Make if you can a rough sketch of the shape you intend it to be. If it has to take the form of a mound, decide where you will make your sheltered bends and where your exposed corners, thinking always of aspect and wind for sun and shelter—decide where it shall rise higher, where shall be a little level plateau, a tiny snug nook, a small moraine—if you want one. When you have got your idea begin to build, using earth and stone at the same time—let the stone support the earth by sloping downwards and inwards so that the soil remains in the little pockets ; the stones must on no account slope outwards, but should be laid naturally, as if they were on a hill-side.   Avoid putting them up on end like almonds in a Christmas pudding.   Don't be disheartened if the result looks very different from the picture you had in your mind's eye ; it will be far better than if there had been no picture.

If you have room to make a rock garden these remarks hold good, only it is still more important to have the ground plan clearly designed—you can adapt it as you work.

Let the paths twist and turn to give the appearance of size. Try and get your levels different, let some parts be low, some broken, some high.   Heighten the high parts with shrubs, grow creeping things on the low ground.

If you are able to have water as well, you are in clover, for you can make the garden charming with the trickle of a rivulet running through, and with the great variety of water plants that can be grown.

Then for the paths : several materials are suitable.   Flat stones set a little apart to allow things to be planted between them ;   old stone edging tiles, if you can get them ;   or the circular mushroom tops on which they lay foundations for stacks in some parts of the country, look delightful ;   if these are not obtainable, broken bits of flags could be used ;   they are quite nice if small enough ;   or else gravel or granite chippings.   If there is room a small stone seat or two is a pleasant addition.

The soil of the rock garden should be good garden soil not heavily manured, or better still virgin soil from under grass or pastureland ; let it be light and rather stony, not heavy ;   work in sand or road grit if it is of a close clay texture.

Now the garden is made you can begin to plant.

First decide where the shrubs are to go ;   the varieties and types depend a great deal on the size and situation of the rock garden. If, as is nicest, it runs off at the back into a wild garden, plant in clumps such things as a variety of brooms, double gorse, veronicas, azaleas, Penzance briars in this wild part ;   in front and among these might grow cistus of various kinds and bulbs.   On the rising ground in the garden itself you might find places for the dwarf juniper, upright and creeping, creeping cupressus, alpine rhododendron (which requires peat), dwarf Japanese maples (which dislike lime), cistus, large heaths, daphne (mezereon), etc. Having planted these turn your attention to smaller plants. If you have anything you particularly prize or that needs a special aspect, plant those first ; then those things that need a special type of place, for instance to fall in a shower over steep rocks, as rock-roses, aubrietia, etc. ; or in a rocky nook as the rosette saxifrages, and erinus ; arenaria balearica insists on a north or north-east side of a rock or it will not grow at all ; tropaeolum speciosum must have its *roots* in a cool place.

If you have no lime in your soil, it is rather a good plan to make one or two beds of old mortar rubble and soil mixed, as many rock plants love lime ; but *don't* plant any heathers, azaleas or rhodo-dendrons on them, these loathe it.

When your favourite plants are in, and are marked with a bit of stick or label if they are small or are things that die down in the winter, fill up your gaps with all your other plants. Saponaria is a boon for the first year in filling up big spaces and growing over the paths and it can be reduced later. Plant it, forget-me-not and any other sturdy and dwarf growing plants between the stones of the paths as they do not mind being trodden on occasionally. Then your bulbs must go in ; mark them with sticks. If you have any creeping roses you will have planted them at the same time as the shrubs, but they are only suitable in large rock gardens.

In the spring fill up any large gaps with annuals ; Venus navel-wort is delightful and will continue to sow itself ; iopsonidum is charming and very dwarf ; brachycome, leptosiphon and phacelia, the last a brilliant blue, are all very useful.

Some seedsmen sell penny packets of seeds of rock plants ; with these you can try a good many varieties with very little expense, but most seedsmen are very expensive for seeds of rock plants.

When planting rock plants between stones see that soil is carefully packed round their roots and always plant as firmly as possible.

Surface rooting things are sometimes difficult to plant ; make

A Water Garden

A Rock Garden

the soil fine and loose, press the plant down, push up the soil round its edge, and, if you can, sprinkle a little soil between the roots on the top and press the plant down, keeping it in its place with a small stone or two if necessary.

If any plants, and many alpines do, dislike damp, they can often be preserved through the winter by putting fine granite chippings round the collar of the plant, which will act as drainage and prevent water lying round the part where the plant touches the ground, which is where it most easily damps off.

Where there is water in the form of either stream or pond, plant water forget-me-not, spireas, iris Kaempferi, and Japanese primula on the damp banks ; in the water itself may be planted iris of various kinds, water-lilies and, in mild climates, arum lilies.   If the bottom of the pond is concrete plant all these in tubs of rich earth placed on the bottom, but if it is a natural pond they can be planted in a large fish-bass of soil ; as they grow the bass will gradually rot and they will root themselves into the bottom of the pond.

NOTES

CHAPTER X

INSECT PESTS AND DISEASES. GRUBS, CATERPILLARS, ETC. MILDEW,
POTATO DISEASE, ETC. A FEW COMMON REMEDIES.

The following are a few of the most ordinary insect pests and plant diseases, and some of the usual remedies.

INSECT PESTS. *American blight.* Scrape the plants infested with it and paint them with potash and quicklime mixed to the consistency of cream, or with a mixture of 5 lbs. lime, 1 lb. sulphur and two gallons of water; this mixture should be heated till the sulphur dissolves; the parts of the tree that cannot be reached should be syringed. A whitewash brush does well to paint with. All loose bark should be removed first and the ground round the stem soaked with soap-suds.

Ants. Pour boiling water or tar into the nest.

Aphis, green fly or black fly. Syringe with some insecticide, or with soft soap and paraffin (soft soap the size of an egg beaten up with hot water, to which add one or two tablespoons of paraffin, mix well, and dilute to two gallons), or with a solution of quassia chips (pour two pints of boiling water on a handful of chips and dilute to two gallons). Indoors, fumigate with nicotine or similar fumigator.

Caterpillars. Hand-picking is the best remedy; infected cabbages may also be washed with soap-suds; a dusting or syringing with hellebore will often cure gooseberries, but it must be applied before the fruit is set, as it is a strong poison.

Celery fly is laid on the leaves of the plants in summer; the grubs are green and should be pinched wherever seen.

Club or finger and toe. This root pest attacks the cabbage tribe; over manuring seems to encourage it. Dressings of charcoal dust, soot, wood ashes, gas lime or salt help to check it.

Currant mite or big bud may be detected by the swollen appearance of the buds on the black currant bushes—they should be picked off and burnt; if badly infected and fruiting poorly destroy the bushes.

Earwigs. Put dirty pots on the tops of sticks and dip them into boiling water every morning to destroy the earwigs.

Grubs. If in the soil, dig in gas lime at the rate of two bushels per 30 square yards; it is best to leave it spread on the ground from November or December to February and then dig it in.

If apples are infected, put a band of cart-grease low down the trunk of the tree to prevent the female grub coming up to lay her eggs; and dig lime into the soil all round.

Onion fly. Dust the ground with charcoal dust, wood ashes and soot. Transplanted onions are rarely affected.

Rosebud maggot. This maggot glues two leaves together. Pinch the maggot hard inside the leaf; if only a few leaves are attacked, pick them off and burn them.

Scale on ferns should be picked off and the fronds sponged with tepid water and painted with paraffin.

Slugs. Dig lime or a slugicide into the soil.

Sprinkle slugicides, soot or lime, or the two last mixed, round young seedlings. Sawdust also prevents them attacking seedlings. Little heaps of bran or half a potato may be put down in the afternoon and examined in the evening and the slugs destroyed. Seed-beds should be examined in the evening by the aid of a lantern and slugs cleared away. Slugicides are best applied with a dredger sold for the purpose.

Worms. Water the lawn, a teaspoon of chloride of lime to two gallons of water, and brush up in the morning.

FUNGOID DISEASES. *Chrysanthemum rust.* Pull off and burn the infected leaves; if a cutting is infected dip it in a solution of sulphide of potassium, half an ounce dissolved in a gallon of hot water.

Mildew on roses and red spider on violets are best treated by syringing or dusting the plants with sulphur. If it attacks apples spray the trees with Bordeaux mixture or sulphur.

Potato disease. Spray the plants as soon as the disease begins to show with Bordeaux mixture, and again a month later.

CHAPTER XI

The following lists are for the convenience of beginners and are not intended to be exhaustive.

ANNUALS.

Aster

Calandrinia Grandiflora (for hot dry borders)

Calliopsis

Centaurea Cyanus Minor (Cornflower)

Clarkia

Eschscholtzia

Godetia

Gypsophila Elegans

Larkspur (Rosy flowering)

Lavatera

Limnanthes

Linum

Love-in-a-mist

Marigold—in variety

Mignonette

Nasturtium—tall and dwarf

Nemesia

Phlox Drummondi

Poppies—especially Shirley

Stocks

Sweet Peas

Sweet Sultan

Tropaeolum Canariensis.

BIENNIALS.

Antirrhinum

Campanulas

Canterbury Bells

Forget-me-not

Foxglove—white and yellow

Polyanthus

Poppy, Iceland

Sweet William

Viola

Wallflower.

6—3

PERENNIALS.

* Can be raised from seed.

Achillea (stands drought)
*Anemone—Fulgens and S. Brigid (His Excellency)
*Aquilegia
*Borage or Anchusa—Dropmore
Campanulas—tall and dwarf
*Carnation
Hardy Chrysanthemum
*Coreopsis
Dahlias
Daisies—tall growing varieties
*Delphinium
Doronicum
Eryngium (blue thistle)
Fuchsia—hardy sorts
Galega (Goat's Beard)
*Geum
*Gypsophila Paniculata, single and double

Helianthus—Mrs Mellish
Heuchera
Iris—English, Spanish and German, especially I. pallida
Japanese Anemones — white and pink
*Linum Perenne
*Lychnis
Montbretia
Paeonies
Phlox
Pinks
Potentillas
Rudbeckia
Saxifrage
*Scabious Caucasica
Statice
*Tree Lupins
Tritoma (red hot poker)

and Michaelmas Daisies, var. Cordifolius, Edith Gibbs, Paniculata, Amellus, Viminius, Dwarf Queen.

ROCK PLANTS.

Perennials.

*Alyssum—Saxatile and Citronum
Anemone—Fulgens, S. Brigid, Hepatica, Sylvestris, Pulsatilla
*Arabis

Arenaria—Balearica (climbing on stone), Montana
Armeria (Thrift)
Aster Alpinus
*Aubrietia—Dr Mules, Leichtlini

Campanulas—Carpatica, Grandi-
flora, Isophylla, Pusilla,
Glomerata, etc.
Cerastium
*Dianthus
*Draba
Epimedium
Erica (Heaths)—Carnea, Men-
ziezia alba, Mediterranea, etc.
Erinus Alpinus
Gentiana
*Helianthemum (Sun Rose)
*Linaria
Lithospernum
*Lychnis
Omphalodes Verna

Phlox—Amoena, Divaricata,
Subulata
*Primula — Rosea, Japonica,
Denticulata, Cashmeriana
*Saponaria Ocymoides
Saxifrage—Androsace, Bick-
ham's Red, Aizoon, Bath-
oniensis, etc.
Sedums
Sempervivum
*Silene
Thymus—silver and golden
Valerian.
Veronica—Rupestris and shrubs
Viola Perenne

Annuals include :

Alyssum Maritimum
Brachycome
Dianthus Laciniatus
Ionopsidium Acaule
Leptosiphon

Linaria
Omphalodes Liniafolia (Venus
Navelwort)
Phacelia.

PLANTS FOR SHADE AND SEMI-SHADE.

Biennial and perennial.

Anthericum (semi-shade)
Crown Imperial
Foxgloves
Funkia
S. John's Wort of various
kinds
Marguerites—yellow and white
Paeony—old common sort
Pansy
Periwinkles — blue, red and
double

Phlox (tall kinds)
Polygonum Orientale
Poppy—Oriental (semi-shade)
Ribbon Grass
Saxifrage—Rosette
Senecio
Silphium
Solomon's Seal
Spireas
Variegated Hemerocallis (Day
Lily).

Annuals (semi-shade).

Alyssum Maritimum

Calendula Officinalis

Clarkia Elegans

Coreopsis

Godetia

Iberis

Limnanthes (does well under trees)

Nasturtium.

List of flowers in a mixed border.

January. Aconites. Iris Reticulata.

February. Aconites. Snowdrops. Primroses. Iris Histrio. Heath.

March. Snowdrops. Crocus. Narcissus.

April. Crocus. Narcissus. Tulip. Wallflower. Forget-me-not. Polyanthus. Hyacinth.

May. Tulip. Hyacinth. Iris. Oriental Poppy. Paeony.

June. Pinks. Iris. Borage. Paeony. Rose. Delphinium. Early annuals. Eschscholtzia.

July. Carnation. Malva. Borage. Delphinium. Rose. Goat's Rue. Lupin. Annuals. English and Spanish Iris. Lilies. Sweet Pea.

August. Montbretia. Carnation. Malva Moschata. Sweet Pea. Lychnis. Bedded out plants. Summer Chrysanthemums. Helianthus. Dahlia. Lilies. Eschscholtzia.

September. Michaelmas Daisy. Helianthus. Dahlia. Rose. Heather. Chrysanthemums. Montbretia. Marguerites. Golden Rod.

October. Michaelmas Daisy. Golden Rod. Cosmia. Dahlia. Marigolds. Chrysanthemums. Commelyna. Alyssum Maritimum.

November. Michaelmas Daisy. Dahlia. Iris Alata.

December. Iris Reticulata. Christmas Roses. A few stray roses and shrubs.

INDEX

Printed in the United States
By Bookmasters